T0298980

This book is an up-to-date introduction to the quantum theory of measurement, a fast developing field of intense current interest to scientists and engineers for its potential high-technology applications.

Although the main principles of the field were elaborated in the 1930s by Bohr, Schrödinger, Heisenberg, von Neumann and Mandelstam, it was not until the 1980s that technology became sufficiently advanced to allow its application in real experiments. Quantum measurement is now central to many ultra-high technology developments, such as "squeezed light," single atom traps, and searches for gravitational radiation. It is also considered to have great promise for computer science and engineering, particularly for its applications in information processing and transfer. The book begins with a brief introduction to the relevant theory and goes on to discuss all aspects of the design of practical quantum measurement systems.

This book is essential reading for all scientists and engineers interested in the potential applications of technology near the quantum limit.

Quantum Measurement

Quantum Measurement

Vladimir B. Braginsky
and Farid Ya. Khalili
Moscow State University

Edited by
Kip S. Thorne
California Institute of Technology

CAMBRIDGE
UNIVERSITY PRESS

Published by the Press Syndicate of the University of Cambridge
The Pitt Building, Trumpington Street, Cambridge CB2 1RP
40 West 20th Street, New York, NY 10011–4211, USA
10 Stamford Road, Oakleigh, Melbourne 3166, Australia

First published 1992
First paperback edition (with corrections) 1995

A catalogue record for this book is available from the British Library

Library of Congress cataloguing in publication data
Braginskii, V. B. (Vladimir Borisovich)
Quantum measurement / Vladimir B. Braginsky and Farid Ya. Khalili
edited by Kip S. Thorne.
    p.   cm.
Includes bibliographical references and index.
ISBN 0-521-41928-X
1. Quantum theory.   2. Physical measurements.   I. Khalili, Farid
Ya.   II. Thorne, Kip S.   III. Title.
QC174.12.B72   1992
530.1'2–dc20   91-40199   CIP

ISBN 0 521 41928 X hardback
ISBN 0 521 48413 8 paperback

Transferred to digital printing 1999

Contents

---

*For an explanation of the asterisks, see the first page of Chapter I.

A quantum phenomenon
is a phenomenon
Only if it is a recorded phenomenon

John Archibald Wheeler

Editor's Foreword

The quantum theory of measurement has been widely regarded, since the 1940s, as an esoteric subject of little relevance to "real" physics. Quite the opposite was true in the 1920s and 1930s when Niels Bohr, John von Neumann, and others were struggling, by means of gedankin experiments, to develop an understanding of how measurements on single quantum objects were to be incorporated into quantum theory. However, the understanding they achieved turned out to be of little use in the quantum mechanical applications that commanded the attention of physicists between 1940 and 1980: the interaction of photons, atomic nuclei, and elementary particles, the theory of masers and lasers, the properties of matter (superfluidity, superconductivity, semiconductors), etc.

In each of these applications, although the phenomena studied were quantum mechanical at heart, the measurements used to probe them were so far removed from the quantum domain that the only feature of the quantum theory of measurement which entered into the experiments was the probability interpretation of squared amplitudes. There was no need to invoke what soon came to be regarded as an esoteric, problematic, dubious "collapse (reduction) of the wave function."

Fundamentally, the reason for this irrelevance of the quantum theory of measurement was technological. The technology of 1940 - 1980 was not capable of making repetitive measurements on a single quantum mechanical system and thereby discovering in a second measurement how a first measurement had affected the system. Instead, the experiments of 1940 - 1980 typically entailed an ensemble of single measurements on a huge number of quantum systems (e.g. muons flying out of the interaction region in a high-energy scattering experiment), with each measurement destroying the quantum system (muon) in the process of measuring its properties.

In the 1980s technology finally began to catch up with the measurement concepts of Bohr, von Neumann, and their 1930s colleagues, and forced modern physicists to elaborate and extend those concepts from the gedankin experiments of the 1930s to the actual experimental situations of the 1980s: The new technology entails repeated measurements on single quantum systems, measurements in which the more "esoteric" features of the quantum theory of measurement are essential.

One example is repeated measurements of the state of a mode of the electromagnetic field in an optical resonator, measurements that are central to the development and use of "squeezed light," "frequency anticorrelated states," "pure number eigenstates," and other non-classical states of light. A second example is repeated measurements of a single atom that is held in an electromagnetic trap. These examples and others hold great promise for fundamental physics experiments (e.g. tests of quantum theory and of relativity theory, and searches for gravitational radiation), and also for practical applications (e.g. information processing and transfer, and high-stability clocks). Correspondingly, we can expect the quantum theory of measurement to become a central tool of scientists and engineers during the coming decades.

Almost all of the textbooks on quantum mechanics are written from the standard viewpoint of the 1940s - 1970s, the viewpoint that has little respect for or interest in the quantum theory of measurement. Correspondingly, it is essential that, until new textbooks take their place, they be supplemented by a small monograph that treats carefully and clearly the quantum theory of measurement and its applications to practical, high-technology experiments. This book is ideal for this purpose.

Despite being far removed from real experiment until recently, the quantum theory of measurement has long been an arena in which theorists carry out intense, emotional battles. Code words such as *many worlds interpretation, collapse of the wave function*, and *irreversibility of measurement* have generated enormous entropy among theorists over the past half century — and have also generated, in the end, some considerable understanding. The reader who wants a guide to the controversies (most of which are over issues of taste and viewpoint rather than over irreconcilable substance) will *not* find it in this small book.* Instead, this book

_____
*For collections of papers that deal with the controversies and the insights they have brought, and with aspects of the quantum theory of measurement that are

focuses on aspects of the theory that by now, in the early 1990s, are rather well understood and fairly noncontroversial, and that, most especially, are central to the real experiments now being made possible by the rapid march of technology.

One can appreciate more fully this book's focus by knowing something about its authors. One of the authors, Vladimir B. Braginsky, is an experimental physicist who has made major contributions to the science of repetitive measurements in the quantum domain. His contributions include the invention of key new ideas for real quantum measurements (e.g. quantum nondemolition measurements and frequency anticorrelated quantum states), and the invention of key new experimental techniques (e.g. the use of "whispering gallery modes" of electromagnetic excitation of dielectric resonators). The second author, Farid Ya. Khalili, is a theorist who has contributed significantly to the modern extensions of the Bohr - von Neumann formal theory of quantum measurements. His contributions have helped extend the idealized Bohr - von Neumann theory into the domain of the practical experimental devices and techniques of real 1990s experiments.

Braginsky's experience with real quantum measurements and his deep physical intuition, when combined with Khalili's deep mathematical insights, gives this book a power that does not exist elsewhere in the pedagogical literature on measurement theory. This combination enables the book to serve simultaneously as an introduction to the formal theory of quantum measurements and as a guide to the physical concepts and the high-sensitivity measuring techniques that make the formal theory relevant to the 1990s and the 21st century.

Kip S. Thorne
California Institute of Technology

---

more esoteric and abstract (some might say more fundamental) than those dealt with in this book, see J. A. Wheeler and W. H. Zurek, editors, *Quantum Theory and Measurement* (Princeton University Press, Princeton, 1983); also H. S. Leff and A. F. Rex, editors, *Maxwell's Demon: Entropy, Information, Computing* (Adam Hilger, Bristol, 1990).

Notation

A	amplitude of oscillations
a, $a^\dagger$	creation and annihilation operators
B	correlation function or correlation matrix
C	capacity of a capacitor
c	speed of light
d	a geometric distance (e.g. distance between plates of a capacitor)
$\boldsymbol{E}$	energy
E	electric field strength
e	charge of the electron
F	force
f	Fourier transform of force
H	magnetic field strength
$\boldsymbol{H}$	Hamiltonian
k	mechanical rigidity
k_B	Boltzmann's constant
$\boldsymbol{L}$	inductance
m	mass
N	number of quanta (chapter XI)
n	number of quanta (except in chapter XI)
	refractive index (chapter XI)
p, P	momentum
q, Q	generalized coordinate
Q	quality factor of an oscillator
q	electric charge
S	spectral density
s	Fourier transform

T	temperature
t	time
U	evolution operator; electrical tension
V	volume
v	speed
w, W	probability; probability density
W	power
x, y, z	spatial Cartesian coordinates
X_1, X_2	quadrature amplitudes
δ	variation of a changing quantity
$\delta(\ )$	Dirac delta function
Δ	uncertainty (standard deviation) of a random quantity
ξ	a dimensionless quantity of order unity
η	flux of the number of photons
λ	wavelength
Ω	reduction operator
ω	angular frequency
ρ	density operator; impedance for a traveling wave; energy density
τ	a time interval
τ^*	relaxation time
ϕ	phase of oscillations; angle
ψ	wave function
χ	generalized susceptibility
$\circ$	symmetrized product of operators: $\hat{Q} \circ \hat{P} \equiv \tfrac{1}{2}(\hat{Q}\hat{P} + \hat{P}\hat{Q})$

Quantities with hats (e.g. $\hat{q}$) are operators

Quantities with tildes (e.g. $\tilde{q}$) are the results of measurements

Spectral densities are normalized to the frequency range $-\infty$ to $+\infty$, so

$$\Delta^2 = \int_{-\infty}^{+\infty} S(\omega) \frac{d\omega}{2\pi}$$

is the squared standard deviation of a quantity with spectral density $S(\omega)$.

I Historical introduction: photons and measurements using photons

In the standard, textbook treatment of quantum mechanics, the contact with experiment is described in terms of probabilities for obtaining this, that, or another result, when identical experiments are performed on a huge number (*ensemble*) of identical objects. Such an ensemble description was adequate during the first half century of quantum theory, because the technology of that era was incapable of making a measurement without destroying or severely changing the measured object (photon, atom, ...). In recent years, however, advances in technology and technique have made possible repetitive quantum measurements on a single quantum object, with each measurement influencing the object only minimally. Such measurements cannot be analyzed using solely the ensemble theory. Additional theoretical concepts are needed. The purpose of this book is to describe the methods and theory of such measurements (as well as the more elementary theory of ensembles of measurements).

This book is written for two types of readers: those who have had only a little previous contact with quantum mechanics, and those who have had much.

Readers of the first type are presumed to understand atomic physics at a qualitative level. The unstarred sections of this book offer such readers an overview of the present state-of-the-art of quantum measurements on single objects and prospects for the future, as well as an elementary overview of the theory of such measurements.

Readers of the second type are presumed to understand quantum mechanics at the level of an advanced undergraduate or first-year graduate course. This book can serve as a supplement to standard textbooks for such a course. The unstarred sections may repeat, in part, material learned from the standard textbooks, but will also extend that material into the domain of measurements on single objects. The starred sections offer the second type of reader an advanced, theoretical understanding of such measurements.

The principal questions addressed by this book are described at the end of this chapter. As background for their description, this chapter presents a short historical excursion into the origin and development of the principal ideas and methods of quantum measurements.

1.1 The discovery of photons

As is well known, quantum physics began with Max Planck's postulate of the discreteness of the energy in a mode of an electromagnetic resonator:[1] the energy comes in discrete quanta, each of which has an energy $E_{quantum}$ proportional to the mode's angular frequency of oscillation ω:

$$E_{quantum} = \hbar\omega . \tag{1.1}$$

The constant $\hbar$ was later named after Planck. This postulate enabled Planck to develop a formal theory that describes accurately the observed spectral distribution of the energy of thermal radiation.

Historically, the next idea was Einstein's:[2] the flux of electromagnetic radiation with frequency ω, like the energy of an electromagnetic oscillator, also consists of discrete quanta, each equal to $\hbar\omega$. This quantum of energy can interact with an object as a whole. Einstein combined this idea with the conservation of energy to derive the following simple formula for the photoelectric effect:

$$\hbar\omega = \frac{1}{2}m_e v^2 + E_{bind} . \tag{1.2}$$

Here ω is the frequency of the incoming electromagnetic radiation, $\frac{1}{2}m_e v^2$ is the kinetic energy of an electron ejected from a metal by the incoming radiation, and E_{bind} is the binding energy of the electron to the metal—a quantity characteristic of the specific metal being used. Einstein's photoelectric formula was confirmed by experiments.

Both of these seminal ideas, Planck's and Einstein's, had a character that was clearly nonclassical; i.e., they could not be derived from Maxwell's equations. These ideas led unavoidably to the conclusion that electromagnetic radiation has not only wavelike properties, but also, simultaneously, particle-like properties. They forced physicists to adapt themselves to the fact that a stream of electromagnetic radiation with frequency ω consists of "seed" portions of energy equal to $\hbar\omega$. Many years later these portions were given the name "photons." Initially this term was used only in the optical region of the electromagnetic spectrum, but now it typically is used throughout the spectrum.

P.N. Lebedev[3] measured the pressure of light (a purely classical effect) several years before the experimental confirmation of Einstein's photoelectric formula (1.2). Lebedev's experiments verified with high confidence the following statement: Any portion E of energy in electromagnetic radiation carries a mechanical momentum P given by

$$P = E/c \, , \qquad (1.3)$$

where c is the speed of propagation of electromagnetic radiation. By comparing the results (1.2) and (1.3) of the photoelectric and light pressure experiments, one is forced to conclude that each photon carries a mechanical momentum

$$P_{\text{photon}} = \frac{\hbar\omega}{c} \, . \qquad (1.4)$$

Experiments on the interference and diffraction of light, when performed with very low light intensities,[4] revealed further that an interference pattern (a classical, pure-wave effect) shows up on a photographic plate only when the number of photons falling on the plate is very large. Each photon in such an experiment is *completely destroyed* (ceases to exist) by interacting with the plate's silver chloride molecules. When the photon is destroyed there appears somewhere on the photographic plate an atom of free silver, which will act as an embryo from which, by photographic developing, a small seed of silver will grow. The silver embryo is much smaller than an electromagnetic wavelength.

This is remarkable. In the interference process (e.g. in the two-slit experiment of Fig. 1.1), the photon must have been influenced by the locations of both slits, since the interference pattern depends on the distance between them. This means that the photon must have occupied a volume larger than the slit separation. On the other hand, when it fell on the photographic plate, the photon must have become localized into the tiny volume of the silver embryo. Later the terms "collapse of the wave function" and "reduction of the wave packet" were used to describe such localization. This collapse or reduction process became one of the key concepts in the quantum theory of measurement.

The wave properties of the photon show up in the fact that the *probability* of collapse at a certain place on the photographic plate (and the accompanying birth of a silver atom there) is proportional to the light

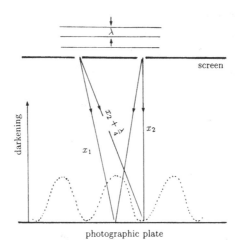

Fig. 1.1 Two-slit interference experiment.

intensity, which is calculated classically using the wave theory. The particle properties show up in the birth of silver atoms, one by one. And the interference pattern appears only after a very large number of photons (an *ensemble* of photons) have fallen on the plate.

Thus, the photon turns out to be neither a particle nor a wave (in the classical sense of these words), but a qualitatively more complicated object; and it behaves like a wave or like a particle depending on the details of the experiment performed.

1.2 The wave and particle properties of photons

In the process of its ''death'' on a photographic plate a photon behaves like a particle, while in the interference process it behaves like a wave. The question naturally arises, can particle properties show up in experiments where the photon does not die? Yes, as we shall see clearly in many examples later in this book. A somewhat fuzzy example, which is useful for delineating the connection between a photon's particle and wave properties, is *localization* of a freely propagating photon. (Localization is a property common to particles and to short wave packets of classical waves.)

The techniques of modern nonlinear optics permit one to prepare very short pulses of light in dielectric waveguides (optical fibers). These pulses can have durations shorter than $\tau \simeq 1 \times 10^{-14}$ sec.[5] Such a pulse in the fiber is localized into a spatial length $\Delta x = c\tau \simeq 3 \times 10^{-4}$ cm, i.e. several optical wavelengths. If the total energy of such an optical pulse is known, then by passing the pulse through several wide-bandwidth absorbers, one can reduce the wave packet's energy to approximately $\hbar\omega$.

Now, the classical theory of waves tells us that the duration τ and bandwidth $\Delta\omega$ of a wave packet are connected by the simple relation

$$\Delta\omega\cdot\tau \geq 1 \ . \tag{1.5}$$

Thus, localization into the spatial interval $\Delta x = c\tau$ must produce an *uncertainty* of the energy. Using equations (1.1) and (1.5), we obtain

$$\Delta E_{\text{photon}}\cdot\tau \geq \hbar \ . \tag{1.6}$$

In other words, whenever an experimenter *prepares* ''short'' photons (photons well localized in space), there inevitably will be a substantial, random unpredictability of the photon energy:

$$E_{\text{photon}} = \hbar\overline{\omega} \pm \frac{\hbar}{2\tau} \ , \tag{1.7}$$

where $\overline{\omega}$ is the mean frequency. However, so long as $\hbar/2\tau$ is small compared to $\hbar\overline{\omega}$, we can be sure that the wave packet contains only one photon.

Instead of the term ''photon localized in space,'' many publications use the phrase ''monophotonic state,'' or ''particle-like wave packet.'' Recently L. Mandel and S. Hong[6] have succeeded in preparing a photon in a monophotonic state, using a scheme proposed by D. Klyshko[7]. The details of their experiment are sketched in Fig. 1.2

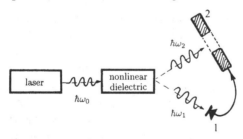

Fig. 1.2 Experimental preparation of a monophotonic state, as carried out by S. Hong and L. Mandel[6] according to a scheme first proposed by D. Klyshko.[7] The photon $\hbar\omega_0$ is split into two photons $\hbar\omega_1$ and $\hbar\omega_2$ in the nonlinear dielectric. The photon $\hbar\omega_1$ is registered by the photodetector 1, which opens the optical gate 2 for a short time τ, permitting the photon $\hbar\omega_2$ to pass. The result is a single photon beyond the gate, localized in a wavepacket of duration τ. This scheme works with a nonideal photodetector (one with quantum efficiency less than unity) just as well as with an ideal one.

Below we shall use the term "photon" in the traditional sense—to denote a quantum of electromagnetic energy of any type: one that is spread out over the volume of a resonator, or one that is localized in a monophotonic state. We shall emphasize localization only when appropriate.

For a monophotonic state the energy uncertainty $\Delta E_{photon} \geq \hbar/\tau$ leads, according to formula (1.4), to the following momentum uncertainty:

$$\Delta P_{photon} \geq \frac{\hbar \Delta \omega}{c} . \qquad (1.8)$$

The uncertainty of the photon's position (relative to the center of the wave packet) is

$$\Delta x = \frac{c\tau}{2} = \frac{c}{2\Delta\omega} . \qquad (1.9)$$

By multiplying equations (1.8) and (1.9), we obtain the fundamental position-momentum uncertainty relation for the monophotonic state:

$$\Delta x \cdot \Delta P \geq \frac{\hbar}{2} . \qquad (1.10)$$

There is no conflict between the above discussion and the fact that Max Planck's original formula $E_{photon} = \hbar\omega$ [Eq. (1.1)] is a precise one. To illustrate this, imagine sending a monophotonic state through a prism (or any other device that decomposes electromagnetic radiation into its spectral components), and onto a photographic plate. The state will produce just one seed of silver on the plate, since it contains just one photon. From the position of this seed relative to the prism, the experimenter can infer the *frequency* of the photon with very high accuracy. In principle, the plate can be replaced by a net of microcalorimeters (devices that measure very accurately the energy of electromagnetic radiation by converting it into some other form of energy). Thereby the experimenter can infer with very high accuracy the photon's *energy* (from the microcalorimeter) and its *frequency* (from the location of the microcalorimeter). The laws of quantum mechanics guarantee that this energy and frequency will be related precisely by Planck's law (1.1). Because this experiment determines, with high precision, the photon's energy and correspondingly its momentum, it must be that in passing through the prism the photon's wave packet gets lengthened by enough to preserve the uncertainty relations $\Delta E_{photon}\tau \geq \hbar$ and $\Delta x \Delta P \geq \hbar/2$. Indeed, one can show that, whenever a monophotonic state is sent through a spectral analyzer, its spatial length increases. This is not peculiar to quantum mechanics; it is true, also, for any classical wave packet.

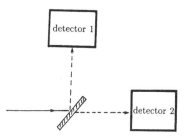

Fig. 1.3 After passing through the beam splitter, the monophotonic state is registered either by detector 1 or by detector 2, and not by both.

To assure ourselves that the photon, despite its wave properties, is also a single, integral particle, we can carry out the following experiment. In the path of a monophotonic state we place a half-silvered mirror (a beam splitter), followed by two photodetectors; cf. Fig. 1.3. Only one of the two detectors will register the photon: detector 1 if the photon passes through the mirror; detector 2 if it is reflected by the mirror. Never, when dealing with one photon, will an experimenter see the monophotonic state's energy get split into two parts and be registered in both detectors (as would be the case for a classical wave packet).

In conclusion, it is worth noting that the particle-wave duality of electromagnetic radiation produces a large number of very interesting effects, which show up in interactions of photons with materials, of photons with phonons, of photons with photons, etc. These effects are now attracting considerable attention among experimenters and theorists, as a result of rapid developments in the state of the art of nonlinear electrodynamics. Despite the popularity of this subject, however, a number of features of photonic interactions remain to be discovered and studied. This book may help to lay foundations for such studies by analyzing and describing various features of the interactions of groups of photons during the measurement process.

1.3 The Heisenberg uncertainty relations

In our historical excursion, we have made a leap of approximately 70 years from the experiments that verified Einstein's photoelectric relation (1.2) to experiments with monophotonic states. Between these experiments, and after Niels Bohr's formulation of his "old quantum theory" postulates, Louis de Broglie hypothesized that ordinary particles have wave properties, and his postulate was confirmed by experiments on the diffraction of electrons (the experiments of Davisson and Germer[8]). In the mid 1920s Werner Heisenberg[9] and Erwin Schrödinger[10] began to formulate mathematically the foundations of the "new" (present-day) quantum mechanics; and in the late 1920s those foundations were completed by Paul Dirac[11] and John von Neumann.[12]

The mathematical formulation of quantum mechanics was far quicker and easier to establish than a full understanding of the *physical essence* of quantum phenomena. Indeed, a full physical understanding is not yet in hand, even now. The most important contribution to our physical understanding was that of Niels Bohr. In his famous discussions with Albert Einstein, Bohr developed the foundations for the physical interpretation of quantum theory—foundations that are now accepted by the majority of physicists.[13,14]

The principal interests of the creators of quantum theory and the quantum theory of measurement were in the microworld. Experimenters during those years focused attention on interactions of *ensembles* of photons with ensembles of nuclei, atoms, and electrons, and on interactions between ensembles of elementary particles. Correspondingly, the creators of quantum theory paid little attention to features of quantum mechanics that lie outside the domain of microworld ensembles; most especially, they largely ignored phenomena associated with measurements of single objects.

As an introduction to such phenomena, we shall recall two thought experiments about measurements of single objects that *were* developed by the creators of quantum theory, as part of their effort to clarify the methodology of quantum measurements: Heisenberg's microscope (end of 1920s), and von Neumann's Doppler speed meter (beginning of 1930s).

The Heisenberg microscope

We shall describe a version of the Heisenberg microscope that is closer to the main contents of this book than Heisenberg's original version.

Suppose that one wishes to measure the position x_1 of a macroscopic body with mass m. To do so, one can attach to the body a stick with diameter less than or of order the wavelength of light. (This can actually be done with modern technology.) If we know in advance the *approximate* position of the mass m, then we can arrange a lens and a photographic plate as shown in Fig. 1.4. The stick must be close to the focal plane of the lens, and the optical amplification factor will be approximately L_2/L_1, where L_1 is the focal length. We can then send in a stream of photons from the side and wait for an *individual* photon to be scattered by the stick, pass through the lens's aperture a, impinge on the photographic plate, and there collapse and produce a small seed of silver.

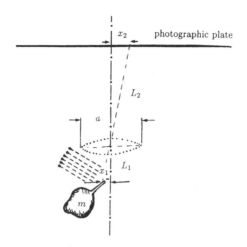

Fig. 1.4 The Heisenberg microscope.

The transverse position x_2 of the silver seed, relative to the lens's optical axis, can be determined to an accuracy much better than an optical wavelength. From x_2 we infer the transverse position $x_1 = -x_2 L_1/L_2$ of the stick, relative to the optical axis. However, we cannot claim that the photon was scattered by the stick at a position precisely equal to this x_1. Because of the photon's wave properties, the scattering may have occurred, with roughly equal probabilities, anywhere within a distance

$$\Delta x_{\text{measure}} \simeq \frac{1}{6}\lambda\frac{L_1}{a} \qquad (1.11)$$

of the location $x_1 = -x_2 L_1/L_2$. Here λ is the photon's wavelength. (This fuzziness in the location of the scattering is well known in optics, for light propagating in the opposite direction to that of Fig. 1.4: A plane wave, when focused by a lens, forms a spot whose size is $\simeq (\lambda/3)(L_1/a)$ — the "Airy spot.") The $\Delta x_{\text{measure}}$ of equation (1.11) is the error in the inferred position x_1 when one observes the scattering of a single photon.

Now, because the photon has a momentum [equation (1.4)], and because it must have passed through the lens's aperture a, it must have given to the stick (and its attached mass m) a random momentum in the x direction, with unknown sign and with magnitude of order

$$\Delta P_{\text{perturb}} \gtrsim \frac{\hbar\omega}{c}\frac{a}{2L_1}, \qquad (1.12)$$

for $a/L_1 \ll 1$. From the product of equations (1.11) and (1.12) we obtain the following form of the Heisenberg uncertainty relation

$$\Delta x_{\text{measure}} \cdot \Delta P_{\text{perturb}} \geq \frac{\hbar}{2} . \qquad (1.13)$$

This thought experiment illustrates the main elements of a quantum measurement:

a)   The extraction of information about the measured quantity x_1 (usually called the *observable*), to within a definite error.

b)   The inevitable perturbation of another quantity P_1 (or quantities) by the measurement process.

c)   An inevitable irreversible process (in our case the death of the photon and birth of the silver seed)—a process that, in fact, is macroscopic in a sense that we shall explore later.

Despite the similar forms of the formulae $\Delta x \cdot \Delta P \geq \hbar/2$ [equation (1.10) for a monophotonic state] and $\Delta x_{\text{measure}} \cdot \Delta P_{\text{perturb}} \geq \hbar/2$ [equation (1.13) for the Heisenberg microscope], their essences are fundamentally different. Equation (1.10) is a fundamental property of the physical state of a quantum object; in it the object's position and momentum have equal footing. According to it, a quantum object, independently of its prehistory (how its state was prepared) cannot have precisely defined values of its position and momentum simultaneously—and this is true in the same sense as a classical wave packet's inability to have simultaneous, precise values of its position and wave number (or frequency). By contrast, equation (1.13) is a fundamental property of the process of measurement; and the uncertainties in the object's position and momentum appear in it in different ways: the position uncertainty is an error in the measurement, the momentum uncertainty is a perturbation given to the object by the measuring process. Nevertheless, both equation (1.10) and equation (1.13) carry the name "Heisenberg uncertainty relation."

In fact, the physical roots of the two relations (1.10) and (1.13) are the same (see chapters II and III and also the beginning of chapter V): From the fact that the state of the *measuring device* has unavoidable uncertainties described by (1.10), it follows that a measurement of the mass's position inevitably perturbs its momentum in the manner of (1.13) (and, similarly, a measurement of the mass's momentum inevitably perturbs its position in a manner analogous to (1.13); see the Doppler speed meter, below). Conversely, the perturbation that inevitably accompanies any measurement prevents one from ever preparing a quantum object in a state with simultaneous, precise values of its position and momentum.

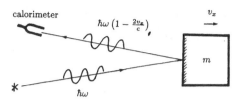

Fig. 1.5 von Neumann's Doppler speed meter.

The Doppler speed meter

The Heisenberg microscope measures position, but not momentum. To demonstrate that the roles of position and momentum can be interchanged, von Neumann[12] formulated the thought experiment shown in Fig. 1.5. A single photon is prepared in a monophotonic state with duration τ (and frequency range $\Delta\omega \simeq 1/\tau$), and propagating in the x direction. The photon's energy is known with an error $\Delta E \simeq \hbar\Delta\omega$ and its position with an error $c/2\Delta\omega$ [cf. equation (1.9)]. If the photon is reflected through an angle $\simeq 180$ degrees by a macroscopic body with mass m and speed v_x, then because of the Doppler effect, the *mean* frequency of the photon will change by the amount

$$\frac{\delta\omega}{\omega} = \frac{2v_x}{c} . \tag{1.14}$$

One can measure the energy of the reflected photon very accurately in an irreversible process. However, the initial energy was known only with a fractional error $\Delta\omega/\omega$, so the mass's speed will be known only with the error

$$(\Delta v_x)_{\text{measure}} = \frac{c}{2}\frac{\Delta\omega}{\omega} = \frac{c}{2\omega\tau} . \tag{1.15}$$

The photon's reflection gives the body a precisely known momentum $2\hbar\omega/c$, but the moment of time when the momentum is transferred is known only up to an uncertainty τ. Correspondingly, the body's position is perturbed by an amount that is at least as uncertain as

$$\Delta x_{\text{perturb}} \geq \frac{2\hbar\omega}{cm}\cdot\frac{\tau}{2} . \tag{1.16}$$

The product of (1.15) and (1.16) is

$$\Delta x_{\text{perturb}}\cdot(\Delta v_x)_{\text{measure}} \geq \frac{\hbar}{2m} , \quad \text{or} \quad \Delta x_{\text{perturb}}\cdot\Delta P_{\text{measure}} \geq \frac{\hbar}{2} . \tag{1.17}$$

Thus, after the change from speed to momentum as the observable, we obtain the standard Heisenberg uncertainty relation, but now with the roles of the position and momentum interchanged.

This thought experiment has the same three important elements as we enumerated when discussing the Heisenberg microscope.

The Heisenberg microscope and the Doppler speed meter are not only thought experiments; they can be carried out in practice, though a practical design for the speed meter was devised only in recent years.

1.4 When do macroscopic objects behave quantum mechanically?

About 20 years ago physical experiments with single *macroscopic* objects became so accurate that experimenters began to face a phenomenon previously regarded as the sole "privilege" of the *microworld*: the quantum mechanically unavoidable back action of the measuring device on the object being measured.[15] This triggered a theoretical exploration of the conditions under which macroscopic objects exhibit quantum mechanical properties. We shall now discuss those conditions.

It should be evident that the more precisely one needs to know the state of a macroscopic object, the more it will behave quantum mechanically. To identify the level of accuracy beyond which quantum effects are decisive, let us analyze a typical measurement procedure: the monitoring of the position of a particle during a time τ.[16] From such a monitoring, one can infer both the particle's mean position, and its speed and thence momentum. We shall confine ourselves to a simple model of such a procedure: a pair of quick measurements of the position, separated by the interval τ. (The full analysis of a detailed experiment is much more sophisticated, but it gives the same results to within a factor of order unity; see chapters V and VI.)

We presume that at time $t = 0$, a "Heisenberg microscope" is used to measure the particle's position with an error $(\Delta x_{measure})_1$, and with a corresponding perturbation of the momentum

$$\Delta P_{perturb} = \frac{\hbar}{2(\Delta x_{measure})_1} . \qquad (1.18)$$

After the time τ, the momentum perturbation will have produced an additional uncertainty in the particle's position

$$\Delta x_{add} = \frac{\Delta P_{perturb}\tau}{m} = \frac{\hbar\tau}{2m(\Delta x_{measure})_1} , \qquad (1.19)$$

which evidently will superpose incoherently on the error of the second measurement $(\Delta x_{measure})_2$.

From the results of the two position measurements, the experimenter infers the momentum of the particle

$$P = m\frac{x_2-x_1}{\tau} . \tag{1.20}$$

The three errors in the position measurements produce the following rms error in the inferred momentum:

$$\Delta P = \frac{m}{\tau}[(\Delta x_{measure})_1^2 + (\Delta x_{measure})_2^2 + (\Delta x_{add})^2]^{1/2} . \tag{1.21}$$

If one wishes the highest possible accuracy for the inferred momentum, one should not make $(\Delta x_{measure})_1$ arbitrarily small, because this will make Δx_{add} arbitrarily large. By combining equations (1.21) and (1.19) one readily sees that it is optimal to choose $(\Delta x_{measure})_1 = (\hbar\tau/2m)^{1/2}$ so that $(\Delta x_{measure})_1 = \Delta x_{add}$. The resulting optimal accuracy for the inferred momentum is

$$\Delta P \geq \Delta P_{SQL} \equiv \sqrt{\frac{\hbar m}{2\tau}} . \tag{1.22}$$

Using the same method it is easy to see that the minimum possible error in the measured mean value $x = \frac{1}{2}(x_1+x_2)$ of the particle's position during time τ is

$$\Delta x_{SQL} = \sqrt{\frac{\hbar\tau}{2m}} . \tag{1.23}$$

The ΔP_{SQL} and Δx_{SQL} of equations (1.22) and (1.23), and other analogous quantities, are now called *standard quantum limits*. Other examples are the standard quantum limit for the quadrature amplitudes of a mechanical oscillator

$$\Delta X_{SQL} = \sqrt{\frac{\hbar}{2m\omega_m}} , \tag{1.24}$$

where m and ω_m are the mass and angular eigenfrequency of the oscillator; and the standard quantum limit for the energy of the oscillator

$$\Delta E_{SQL} = \sqrt{\hbar\omega_m E} , \tag{1.25}$$

where E is the oscillator's mean energy. For details see section 4.1.

It should be evident that a macroscopic object begins to behave quantum mechanically when the measurements that probe it reach accuracies of the order of the standard quantum limit. To achieve such accuracies, one must ensure that all unpredictable, fluctuating classical forces that act on the object are sufficiently small. This leads to additional criteria that must be satisfied if the object is to behave quantum mechanically.

The most fundamental of these additional criteria are associated with fluctuating thermal forces. These will be the relevant criteria whenever industrial, seismic, acoustical, and other forces have been excluded from the experiment using filters, screens, isolators, etc.

Consider, as an example, a harmonic oscillator. For an oscillator an often cited criterion for thermal fluctuations to be small compared to quantum fluctuations (and therefore for the oscillator to behave quantum mechanically) is

$$k_B T \leq \frac{\hbar \omega}{2} . \tag{1.26}$$

Here k_B is Boltzmann's constant and T is the temperature of the environment in which the oscillator lives. This is the relevant criterion if one is interested only in the rms fluctuations of the oscillator, averaged over times τ long compared to the oscillator's relaxation time τ^*. However, one often deals with measurements of the oscillator that last for a time $\tau \ll \tau^*$. During such a short time the oscillator's thermalized environment randomly exchanges with the oscillator an energy far smaller than $k_B T$, and correspondingly, equation (1.26) is not the correct criterion for quantum behavior. It can be shown, by analyzing the random forces ("Nyquist forces") exerted by the oscillator's environment, that the random, thermally induced change of the oscillator's amplitude during the time $\tau \ll \tau^*$ has an rms value

$$\Delta X_{th} = \sqrt{\frac{k_B T \tau}{m \omega^2 \tau^*}} . \tag{1.27}$$

By comparing this thermal perturbation with the corresponding standard quantum limit (1.24), we infer that the oscillator can behave quantum mechanically only if

$$\frac{2 k_B T \tau}{\omega \tau^*} \leq \hbar , \tag{1.28}$$

a criterion satisfied at much higher temperatures than (1.26). Indeed, for relaxation times τ^* that have been achieved by experimenters, both in mechanical and in electromagnetic resonators, condition (1.28) can easily be satisfied at liquid helium temperatures.[17,18]

As a second example, consider a "free" particle (one on which the only significant forces are those of the measuring device, and the frictional force due to its thermalized environment). If, as for the oscillator, $\tau \ll \tau^*$, then the random displacement of the particle during time τ has the rms value

$$\Delta x_{th} \simeq \sqrt{\frac{k_B T \tau^3}{m \tau^*}} \; . \tag{1.29}$$

From this formula and equation (1.23) we infer the following criterion for the particle to behave quantum mechanically:

$$\frac{2 k_B T \tau^2}{\tau^*} \leq \hbar \; . \tag{1.30}$$

For a measurement time $\tau \lesssim 1 \times 10^{-3}$ sec, and for a relaxation time $\tau^* = 4 \times 10^7$ sec (which has been achieved when the particle is supported against falling by a thin fiber, but moves essentially freely horizontally), criterion (1.30) is satisfied even at room temperature, $T = 300$ K.

In conclusion, it is important to notice that the above thermal criteria, equations (1.28), (1.30), are relevant only when the goal is to see the measured object behave quantum mechanically. For experiments with precisions much better than the standard quantum limit, thermal effects must be kept smaller than (1.28), (1.30), and their control may be much harder. For example, an attractive goal is to monitor the energy of a resonator with an accuracy equal to one quantum, $\Delta E \simeq \hbar \omega$. For such an experiment, the random change of energy due to thermal fluctuations during the measurement, $\Delta E = \sqrt{\hbar \omega E \tau / \tau^*}$, must not exceed $\hbar \omega$. From this, and by expressing the resonator's mean energy as $E = (n + \frac{1}{2}) \hbar \omega$ where n is the mean number of quanta in the oscillator, we infer that, instead of criterion (1.28), the following criterion must be satisfied:

$$(n + \frac{1}{2}) \frac{k_B T \tau}{\omega \tau^*} \leq \hbar \; . \tag{1.31}$$

In other words, to monitor the resonator's energy at the level $\hbar \omega$ requires a temperature smaller by a factor $2/(n + \frac{1}{2})$ than to simply see that the resonator is quantum mechanical.

1.5 Overview of this book

Chapters II and III of this book present the basic concepts of the quantum theory of measurement as formulated in the 1920s and 30s by Bohr, Schrödinger, Heisenberg, von Neumann, and Mandelstam.

Chapter II discusses the key concept of the quantum theory of measurement, the reduction of the quantum state, and von Neumann's mathematical apparatus based on this concept. This chapter also presents some aspects of the theory of "quasimeasurements," which was formulated at the end of the 1960s and beginning of the 70s as an extension of von Neumann's ideas.

Chapter III deals with the theory of indirect quantum measurements. This theory permits one to combine the physical properties of the measuring device with the procedure of measurement, and thereby produce a detailed description of the measurement. This description of quantum measurements was proposed already in the 1930s by Leonid Mandelstam, but it has been fully developed only recently, as a result of the development of technology for carrying out quantum measurements on macroscopic systems.

Subsequent chapters deal with issues that were not discussed by the creators of the quantum theory of measurement. Among those issues are the following:

1. How must the measuring device be designed to ensure that the perturbations it exerts on the measured object do not interfere with the measurement, or their interference is minimized (chapter IV)?

2. How does the measured object behave if the measurement lasts for a finite time, and if the experimenter is extracting information continuously during the measurement (chapters V—VII)? In particular, chapter VII analyzes continuous, nonlinear measurements, in which there can occur highly nonclassical effects that have attracted much recent attention.

3. How must a measurement be designed in order to detect with the highest possible confidence the action of a *classical* force on a *quantum* object, and what are the fundamental limits that quantum mechanics enforces in this case (chapters VIII—X)? In chapter X, in particular, the ultimate possible sensitivities of real detection systems are computed.

4. What can be learned about the state of one group of photons, using its interaction via a nonlinear device with another group of photons (chapter XI)?

5. What accuracies can be achieved in experiments such as the Doppler speed meter if, instead of a single photon, a group of photons is used, and how must the group of photons be prepared in order to achieve the optimal accuracies (chapter XII)?

II   The main principles of quantum mechanics

2.1 The wave function

Before beginning to describe the quantum theory of measurement, we shall discuss briefly the main concepts of quantum mechanics. Readers familiar with quantum mechanics may skip this section and go on to section 2.2.

Consider, as a pedagogical aid, a classical particle that can move only along the x direction. For example, it might be a bead on an infinite needle. If we know at any moment of time the values of two numbers, the particle's position and momentum, then we can compute from them all other features of the particle's motion: the forces acting on it, its resulting acceleration, etc. In other words, this pair of numbers, the position and momentum, completely define the state of the classical particle at the chosen moment of time. Similarly, for three-dimensional motion the state is defined by six numbers: the three spatial coordinates x, y, z and the three corresponding components of the momentum p_x, p_y, p_z.

A quantum object, with its dual particle-like and wave-like properties, is much more complex than a classical particle. Interferometric experiments (chapter I) show that the object can be spread out over some region of space, or for the bead on a needle, over an interval of x. As a result, to define the state of a quantum object one needs a function of its coordinates, called the "wave function" and denoted ψ. In the one-dimensional case, ψ at a chosen moment of time depends on the one coordinate x, and in the three-dimensional case, on the three coordinates

x, y, z.

To understand the physical meaning of the wave function, one must study two issues:

First, how is the wave function related to the values of physical quantities such as the object's position and momentum? In other words, what can be said about the results of measurements of a chosen physical quantity if we know that the quantum object is in a state with some given wave function?

Second, what is the law that describes the change of the wave function with time? In other words, what is the quantum analog of the object's classical equations of motion?

The first of these issues is a central theme of this book, and we shall begin to discuss it in the next section. The remainder of this section is a brief discussion of the second issue.

The equation of time evolution for the wave function was first formulated by Erwin Schrödinger and is named after him. In the case of the one-dimensional motion of a particle, this Schrödinger equation has the following form:

$$i\hbar\frac{\partial\psi(x,t)}{\partial t} = -\frac{\hbar^2}{2m}\frac{\partial^2\psi(x,t)}{\partial x^2} + V(x)\psi(x,t) ,$$

where m is the mass of the particle and $V(x)$ is the potential energy for whatever external forces act on the particle (e.g. $V \equiv 0$ for a free particle and $V = \frac{1}{2}Kx^2$ for a particle attached to a spring with rigidity K).

The general form of the Schrödinger equation is the following:

$$i\hbar\frac{\partial\psi}{\partial t} = \hat{H}\psi , \qquad (2.1)$$

where $\hat{H}$ is the so-called "Hamiltonian operator." Simplifying a little, one can regard $\hat{H}$ as the total energy of a classical version of the object, expressed as a function of the coordinates and momentum x, y, z, p_x, p_y, p_z, with p_x replaced by its "operator form" $\hat{p}_x \equiv (\hbar/i)\partial/\partial x$ and similarly for p_y and p_z. For example, if the object is a one-dimensional harmonic oscillator, then

$$\hat{H} = -\frac{\hbar^2}{2m}\frac{\partial^2}{\partial x^2} + \frac{Kx^2}{2} .$$

By expressing $\hat{H}$ in the Schrödinger equation (2.1) in terms of partial derivatives and solving the resulting differential equation, one learns how the wave function changes with time, i.e. how it evolves. Most standard textbooks dwell at length on exact solutions of the Schrödinger equation for important, simple objects (free particle, oscillator, rotator, hydrogen atom, ...), and on approximate methods of solving the Schrödinger equation. We refer the reader to the textbooks for these standard issues.

It is appropriate to note that in many cases the solution of the Schrödinger equation takes the form of a wave propagating through space; hence, the term "wave function."

2.2 Probabilistic interpretation of the wave function

So long as one avoids the issue of measurements, one can regard the wave function ψ for a quantum particle as an ordinary classical wave that occupies a certain region V of space. Radical differences from a classical wave show up, however, when experimenters try to "chip off" a part of the wave—i.e., when they insert a calorimeter-type measuring device in some piece of the region V, and with it try to absorb the portion of the wave's energy that lies there.

If the wave were classical, then the calorimeter could absorb all the energy that falls in it, while leaving all the wave energy outside it unaffected. Not so for the wave function of a quantum particle. Inside the calorimeter there must be either all of the particle, or none of it. However, one cannot predict precisely whether the particle is in the calorimeter or not. One can only give a probability for it being there.

That probability is determined by the square of the modulus of the particle's wave function. More specifically, according to one of the fundamental postulates of quantum physics, the so-called "probabilistic interpretation of the wave function," the probability that the particle will be found between x_1 and x_2 is given by

$$W(x_1 < x < x_2) = \int_{x_1}^{x_2} |\psi(x)|^2 dx . \qquad (2.2)$$

In other words, the square of the modulus of the wave function is equal to the probability density for the particle's x coordinate. It is important to emphasize that this statement is a postulate which does not follow from the other postulates of quantum physics, but rather is based only on the results of experiment.

To describe fully the state of the particle, one needs to give information not only about its position, but also about its momentum. To understand how the wave function contains this information, let us return to the specific case of a photon. Equations (1.1) and (1.3) tell us that the photon's momentum is coupled to its frequency by

$$P = \frac{\hbar\omega}{c} .$$

This formula can be rewritten in an alternative form:

$$P = \hbar k \ ,$$

where k is the photon's wave number, $k = 2\pi/\lambda$ with λ the wavelength. Notice that in this relation $P = \hbar k$ the speed of light c has disappeared. There is no longer any explicit trace of the fact that we are dealing with a photon.

In fact, in quantum mechanics it is postulated that the relation $P = \hbar k$ is valid not only for photons, but for any kind of particle. This statement was first made by Louis de Broglie,[19] and it was confirmed by experiments on the diffraction and interference of elementary particles.

One cannot define a precise wave number k for a wave packet of finite spatial size. Instead, one must speak about the spectral distribution of its values. The spectral distribution is expressible in terms of the Fourier transform of the wave function,

$$\phi(k) = \frac{1}{\sqrt{2\pi}} \int\limits_{-\infty}^{+\infty} \psi(x)e^{-ikx}dx \ .$$

In the case of a classical wave, the square of the modulus of $\phi(k)$, i.e. $|\phi(k)|^2$, describes the spectral distribution of the wave's energy. Similarly, according to de Broglie's postulate, for a quantum wave function the quantity $|\phi(k)|^2$ determines the probability density for the particle's momentum. More specifically, the probability that a precise measurement of the momentum will give a result in the range $P_1 < P < P_2$ is given by

$$W(P_1 < P < P_2) = \int\limits_{P_1}^{P_2} |\psi_P(P)|^2 dP \ ,$$

where

$$\psi_P(P) = \frac{1}{\sqrt{2\pi\hbar}} \int\limits_{-\infty}^{+\infty} \psi(x)e^{-iPx/\hbar}dx \ . \tag{2.3}$$

This ψ_P is called the momentum representation of the wave function.

It is well known that the more "narrow" is any wave function (i.e. the narrower is the range of x over which it is nonzero), the wider is its spectral distribution. Correspondingly, the smaller is the uncertainty of the position of a quantum object (the more narrow is its wave function ψ in the coordinate representation), the larger will be the uncertainty of its momentum. A rigorous analysis based on equation (2.3) shows that the product of the uncertainties of the position and the momentum always satisfies the inequality

$$\Delta x \cdot \Delta P \geq \frac{\hbar}{2} .$$

Thus, the Heisenberg uncertainty relation enters into the mathematical apparatus of quantum mechanics in a natural way.

2.3 Single measurements and ensembles of measurements

What information can one obtain about the wave function ψ of a quantum object from a single measurement? In the absence of any *a priori* information (information before the measurement), practically nothing. For example, if the object is a particle and the measurement is the detection of the particle at some specific location x on a photographic plate, one can only say that x is in the domain where ψ was nonzero.

To learn the shape of $|\psi|^2$, one must carry out an ensemble of such detection measurements (i.e. a large number of such measurements). This is the strategy of traditional quantum measurements, e.g. those of atomic physics in which one studies the radiation emitted by an atom. These experiments use a very large number of particles, all in nearly identical quantum states; and by detecting each of these particles, one infers the form of the quantum state's wave function with any desired accuracy.

This book deals not with such ensembles of measurements, but rather with measurements of a single quantum object. The effectiveness of such measurements depends substantially on one's *a priori* information about the state of the measured object.

As an example, consider an experiment to detect the action of an external force on an oscillator. (We shall call it the *probe* oscillator, since it is being used to probe experimentally the external force.) The minimum information that one might expect to have about the initial state of the probe oscillator is its mean thermal energy $k_B T$, where k_B is Boltzmann's constant and T is its temperature. An unknown, impulsive force F acts on the probe oscillator for a time τ_F short compared to the oscillator's period, giving it a momentum of order $F\tau_F$. The experimenter then makes a single measurement of the oscillator's energy and from that measurement seeks to determine whether the force actually acted and how strong it was. Clearly, the measurement can yield the desired information only if the momentum transferred to the oscillator, $F\tau_F$, exceeds the oscillator's initial rms thermal momentum P_{th}, which is given by

$$\frac{P_{th}^2}{2m} = \frac{k_B T}{2} ,$$

where m is the oscillator's mass. In other words, the minimum detectable force is given by

$$F \tau_F \geq \sqrt{k_B T m} \ .$$

By decreasing the temperature, one can increase the sensitivity of this experiment. If the oscillator's internal fluctuations were governed by classical mechanics, then in the limit of vanishing temperature, the fluctuations would disappear completely, and the experiment would achieve an arbitrarily high sensitivity. However, in the real quantum world, the Heisenberg uncertainty relations prevent the internal fluctuations from disappearing in the limit of zero temperature.

What is the rms level of the minimum allowed fluctuations? Stated differently, what is the smallest mean energy

$$\langle E \rangle = \frac{\langle P^2 \rangle}{2m} + \frac{K \langle x^2 \rangle}{2}$$

that the probe oscillator can have and still satisfy the Heisenberg uncertainty relation

$$\Delta x \cdot \Delta P \geq \frac{\hbar}{2} \ .$$

The mean of the squared position can be written as a sum of the square of the mean and the variance

$$\langle x^2 \rangle = \langle x \rangle^2 + (\Delta x)^2 \ ,$$

and similarly for the momentum. Thus,

$$\langle E \rangle = \frac{\langle P \rangle^2}{2m} + \frac{K \langle x \rangle^2}{2} + \frac{(\Delta P)^2}{2m} + \frac{K (\Delta x)^2}{2} \ .$$

The mean energy is evidently minimized if the mean position and momentum vanish,

$$\langle E \rangle \geq \frac{(\Delta P)^2}{2m} + \frac{K (\Delta x)^2}{2} \ .$$

In view of the uncertainty relation we can rewrite this as

$$\langle E \rangle \geq \frac{1}{2m} \left[\frac{\hbar}{2\Delta x} \right]^2 + \frac{K (\Delta x)^2}{2} \ ,$$

which is minimized if

$$\Delta x = \sqrt{\frac{\hbar}{2m \omega}} \ , \tag{2.4}$$

where $\omega = \sqrt{K/m}$ is the oscillator's eigenfrequency. The result is the following minimum possible value for the oscillator's mean energy:

$$\langle E \rangle_{min} \geq \frac{\hbar\omega}{2} \, . \tag{2.5}$$

Thus, even when the temperature vanishes the oscillator has some residual energy. This energy is associated with so-called zero-point fluctuations in the oscillator's position and momentum. In these zero-point fluctuations the rms deviation of the position from equilibrium is given by (2.4), and correspondingly (as one infers from the uncertainty principle), the rms momentum is

$$\Delta P = \sqrt{\frac{\hbar\omega m}{2}} \, .$$

These zero-point fluctuations place a limit on the sensitivity of the single-measurement experiment to detect the weak force F, discussed above: In the limit of vanishing temperature, the sensitivity $F\tau_F \geq \sqrt{k_B Tm}$ must be replaced by

$$F\tau_F \geq \sqrt{\frac{\hbar\omega m}{2}} \, .$$

This is the *standard quantum limit* for measurement of an impulsive force using a probe oscillator.

By improving the *a priori* information about the probe oscillator's initial state, one can obtain a better sensitivity than this standard quantum limit. For example, one can carefully adjust the oscillator's wave function into an initial state for which the momentum uncertainty at the time the force acts is smaller than the zero-point uncertainty $\sqrt{\hbar\omega m/2}$; and one can then make a single measurement of the oscillator's momentum immediately after the force acts. This strategy is compatible with the uncertainty relation, since the initial state automatically will have a position uncertainty that is larger than the zero-point value. In chapter VIII we shall see how such an experiment can be carried out in practice.

The process of adjusting the probe's initial wave function is called *preparation of the initial quantum state*, or simply *state preparation*. Often experimenters carry out repetitive measurements on the same object. In such an experiment, each measurement in effect prepares the state of the object for the next measurement. We shall meet many examples later in this book.

The above discussion teaches us several things: (i) To carry out useful measurements on a single quantum object, one must be able to prepare the object in some desired quantum state. (ii) The quantum theory of measurement must not only answer the question of what are the possible results of a measurement when the measured object has a given wave function (a question answered by the wave function's probability interpretation); it must also answer the question of what is the state of the measured object after the measurement is finished. This second question will be addressed in the next section.

2.4 Reduction of a quantum state

One of the key concepts in the description of the measurement process is the reduction of the quantum state of the measured object. To elucidate the essence of this concept, we begin with a simple example of state reduction in classical physics.

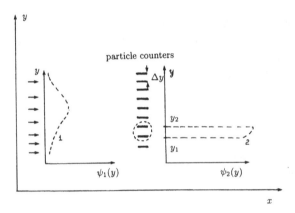

Fig. 2.1 An illustration of the reduction of the wave function.

Consider a beam of classical particles that fly from left to right, as shown in Fig. 2.1. The density of particles along the beam's transverse profile is given by curve 1. This curve, which is the *a priori* probability distribution for the particles' y coordinate, is determined by the particles' source and by the details of their motion in the beam. The experimenter measures the y coordinate of each particle using a sieve. Each cell of the sieve contains a measuring device that records the particle's passage without disturbing it. The size Δy of each cell defines the precision of the measurement.

Suppose that the particles in the beam arrive at widely separated times, and that we focus on a specific particle which is arriving during the time of a specific measurement. Suppose that the measuring apparatus reports that the particle has flown through the cell that is circled in Fig. 2.1. Then after the measurement our knowledge of the particle's y coordinate will be described by the probability distribution of curve 2. The squared area under curve 2, like that under curve 1, must be unity, since we know that there was just one particle present during the time of the experiment, and we know that it passed somewhere through the circled cell. If the width Δy of the cell is less than the initial dispersion in y, i.e. less than the width of the *a priori* probability distribution (as must be the case if the experiment is to be a useful one), then the *a posteriori* probability distribution (curve 2) will be narrower than the *a priori* one (curve 1). This process of narrowing of the probability distribution for the measured quantity, which accompanies the extraction of experimental information about it, is called *reduction*.

What new features does quantum mechanics bring to this example? To answer this question we must go beyond the probability interpretation of the wave function. The probability interpretation tells us only the probability of finding the particle in each cell; it does not say anything about the state of the particle after the apparatus registers it as having passed through the circled cell.

To identify the new quantum mechanical features, we shall now regard the particle as quantum mechanical rather than classical, and we shall suppose that the sieve's apparatus makes two measurements of the particle's location y, separated by a time interval so short that the particle can move only a tiny fraction of a cell size between the measurements. (This discussion is based on one that John von Neumann used as an aid in formulating his postulate of reduction of a quantum state.[12])

The result of the first measurement is predicted by the probability interpretation of the initial wave function $\psi(y)$: it is a random cell drawn from the probability distribution $|\psi(y)|^2 \Delta y$. The result of the second measurement, however, cannot be predicted until we understand the effect of the first measurement on the particle's wave function. That understanding is the goal of this discussion.

One can imagine two different possible outcomes for the second measurement: the outcome might be a random cell drawn from the same probability distribution $|\psi(y)|^2 \Delta y$ as governed the first measurement; or it might be always the same cell (the circled one) as was identified in the first measurement. To learn the answer, we must turn to experiment. Real experiments have revealed that Nature chooses the second possibility: The first measurement leaves the particle in a state with y confined to the cell in which the particle was found; this is von Neumann's reduction of the wave function. (The experiments that von Neumann used to deduce his concept of reduction were those of A. Compton and D. Simons on the scattering of light by electrons. In these experiments the position of the electron was inferred by two different methods separated by a short time interval.[20])

Notice that this outcome for the second measurement is the same in quantum mechanics as in classical mechanics; cf. Fig. 2.1. There is a fundamental quantum/classical difference, however, if the second measurement examines a different quantity from the first measurement. To understand this difference, let us return to the example of a position measurement made with the Heisenberg microscope. Immediately after the collision of the photon with the measured particle, the particle's position is the same as that revealed by the measurement (to within the measurement accuracy). However, the particle's momentum is perturbed by the measurement, and the higher is the measurement accuracy, the stronger is the perturbation; see equation (1.13). The same thing happens when any other physical quantity is measured: The process of measurement perturbs that quantity which is canonically conjugate to the measured one. (Recall from classical physics that each physical degree of freedom of a system is described by a canonically conjugate pair of quantities, and to define completely the state of that degree of freedom, one must specify the values of its two canonically conjugate quantities. Familiar examples are the coordinate and momentum of a free particle, the angle of rotation and angular momentum of a rigid rotator, and the energy and phase of an oscillator.)

The physical source of the perturbation is evidently the measuring device. However, the minimum allowed perturbation does not depend in any way on the details of the measuring device; rather, it is determined solely by the Heisenberg uncertainty relations. In this sense, the perturbation is fundamental.

Since the minimum perturbation follows from the uncertainty relations, its magnitude can be inferred without paying any attention to the back action of the measuring device on the measured object. From this one sees clearly that there is a deep connection between the existence of the perturbation and the particle-wave duality of quantum objects.

To obtain a deeper understanding of the minimum perturbation, let us return to the example of Fig. 2.1. As has been mentioned occasionally already, the wave function of a quantum particle diffracts off obstacles in the same way as does a classical wave with wavelength

$$\lambda = \frac{2\pi\hbar}{P} \; .$$

Here P is the particle's total momentum. Thus, after the particle passes through the circled cell of Fig. 2.1, its wave function must diverge, relative to the line of propagation of the original beam, with an angle

$$\phi \simeq \frac{\lambda}{\Delta y} \; .$$

This is equivalent to the creation of an additional uncertainty in the particle's transverse (y) momentum given by

$$\Delta P_y = \frac{1}{2}\phi P \simeq \frac{\hbar}{\Delta y}.$$

Once more we have obtained the Heisenberg uncertainty relation.

Thus, the reduction of the state is a sophisticated phenomenon in which are intermixed the pure dynamical process of the interaction between the measuring device and the measured object, and the pure information-acquisition process by which the experimenter obtains new knowledge about the object.

The principal conclusions of this section can be summarized by three statements—statements which, like this section, deal with *ideal* measurements in which all non-fundamental sources of error have been excluded.

I. According to the principle of the probability interpretation of the wave function, the probability to obtain this or that result in a measurement is determined by the square of the modulus of the wave function, in the mathematical representation of the measured quantity (position, momentum, or ...).

II. The measurement process leaves the measured quantity unperturbed. There is only a change in our knowledge about it, a change reflected in the narrowing of the probability distribution from a wider, *a priori* one to a narrower, *a posteriori* one.

III. The quantum measurement is accompanied by an inevitable perturbation of the quantity that is canonically conjugate to the measured one. The magnitude of the minimum perturbation is given by the Heisenberg uncertainty relation.

These statements are the essence of the fundamental postulate of the quantum theory of measurement: John von Neumann's postulate of reduction. A more rigorous description of this postulate will be given in the next section.

2.5* von Neumann's postulate of reduction

The rigorous description of what happens to a quantum object when it is measured was first formulated by John von Neumann, in the form of a postulate. This postulate was later named after him. We shall discuss, initially, the simplest case (as was done by von Neumann himself[12]) of an exact measurement of an observable whose operator has a discrete spectrum of eigenvalues. Let q be the observable, $\hat{q}$ its corresponding operator, q_n its eigenvalues (with n an integer), and $| q_n >$ its corresponding eigenstates. Then von Neumann's postulate of reduction says:

1. The result of the measurement of the observable q, which is in a state with the density operator $\hat{\rho}_{\text{init}}$, will be one of its eigenvalues q_n, and the probability of obtaining the various eigenvalues is

$$w_n = \langle q_n | \hat{\rho}_{\text{init}} | q_n \rangle .$$ (2.6)

2. After the measurement, the object will be in the state $| q_n \rangle$ that corresponds to the result of the measurement, q_n.

From this postulate it follows that, if two measurements of q are made, one immediately after the other, then the second measurement must give the same result as the first. This is sometimes called the principle of repeatability of the results of a measurement. To derive this principle of repeatability from the reduction principle, one need only note that the density operator just after the first measurement (and thus also just before the second) is $\hat{\rho}_{\text{init}} = | q_n \rangle \langle q_n |$; and correspondingly (because of the orthogonality of the eigenstates of q), equation (2.6) predicts unit probability $w_n = 1$ for the second measurement to give the same result q_n as the first, and zero probability for any other result.

It is important to note that in many *real* measurements the final state of the object differs substantially from $| q_n \rangle$. A simple example is a measurement of the number of photons in an electromagnetic resonator by absorbing them into a photodetector; in this example, the measurement leaves the resonator devoid of photons. This photon destruction is an example of the fact that a real measuring device can produce additional back-action effects on the measured object, above and beyond the fundamental back action demanded by the von Neumann postulate. Such additional back actions are not fundamental. In principle, they can be avoided. For example, in the case of a resonator one can measure the photons without absorption. von Neumann's postulate of reduction is restricted to ideal measurements in which all such avoidable back actions are excluded.

The von Neumann postulate as formulated above is restricted to measured objects with one degree of freedom. The presence of two or more degrees of freedom can change the character of the measurement substantially. As an example, consider the following version of a famous thought experiment due to Einstein, Podolsky, and Rosen:[21]

A particle with zero spin decays into two with spin ½, and then the projection of the spin of one of the particles on a previously chosen axis (the z-axis) is measured. According to the von Neumann postulate, the result of the measurement must be either $+½$ or $-½$, with equal probabilities for the case of a symmetrical decay.

What can be said about the z-component of the spin of the second particle? In the above formulation, the postulate of reduction leaves this question open. However, the conservation law for total angular momentum requires that the vector sum of the spins of the two particles must vanish, and therefore, if the first particle is measured to have a z-component of spin equal to $+\frac{1}{2}$, the second inevitably must have its z-component equal to $-\frac{1}{2}$. Experiments along these lines have shown that, indeed, there is an unambiguous coupling between the z-components of the two particles. This coupling enforces angular momentum conservation and causes the measurement of the z-component of spin for the first particle to reduce the state not only of the spin of the first particle, but also of the second particle.

The following formulation of the postulate of reduction takes account of this effect. (All notation is the same as above.)

1. The result of the measurement of the observable q, which is in a state with the density operator $\hat{\rho}_{\text{init}}$, will be one of its eigenvalues q_n, and the probability of obtaining the various eigenvalues is

$$w_n = \text{Tr}(|q_n><q_n|\hat{\rho}_{\text{init}}) . \tag{2.7}$$

2. After the measurement, the object will be in the state

$$\hat{\rho}_n = \frac{1}{w_n}|q_n><q_n|\hat{\rho}_{\text{init}}|q_n><q_n| . \tag{2.8}$$

It is easy to see that for an object with just one degree of freedom, this formulation is identical to the previous one. For our example of two spin $\frac{1}{2}$ particles formed by decay of a spin 0 one, the decay puts the two particles into the state

$$|\psi_{\text{init}}> = \frac{1}{\sqrt{2}}\left[|+\frac{1}{2}|-\frac{1}{2}> + |-\frac{1}{2}|+\frac{1}{2}>\right] , \tag{2.9}$$

where $|+\frac{1}{2}|-\frac{1}{2}>$ denotes the state in which the first and second particles have z-components of spin $+\frac{1}{2}$ and $-\frac{1}{2}$, and similarly for $|-\frac{1}{2}|+\frac{1}{2}>$. For each particle the probabilities of the two spin orientations in $|\psi_{\text{init}}>$ are equal, and thus the z-components of spin are totally indeterminate.

The measurement of the first particle, as governed by the reduction equation (2.7), will transform the particles' spins into either the state $|+\frac{1}{2}|-\frac{1}{2}>$ or $|-\frac{1}{2}|+\frac{1}{2}>$, depending on the result of the measurement. (The details of the simple calculation are omitted.) Thus, the measurement of one particle drives both particles into states with precisely defined spin projections in just the manner predicted by the above discussion of angular momentum conservation.

2.6* Orthogonal measurements

We turn attention, now, from idealized, exact measurements, to the development of a formalism that will help us to handle more nearly realistic, approximate measurements. That formalism will reach fruition in the next section.

We begin by rediscussing the sieve-apparatus measurement of the y position of a particle (section 2.4 and Fig. 2.1 above). If we denote by $\psi(y)$ the particle's wave function just before it reaches the sieve and is measured, and we denote by j the cell in which the particle is registered, then immediately after the measurement the wave function will be

$$\psi_j(y) = \begin{cases} (1/N)\psi(y) & \text{if } y_j \leq y < y_{j+1} \\ 0 & \text{if } y < y_j \text{ or } y \geq y_{j+1} \end{cases}, \quad (2.10)$$

where y_j and y_{j+1} are the edges of cell j, and

$$N = \left[\int_{y_j}^{y_{j+1}} |\psi(y)|^2 \, dy \right]^{1/2}$$

is the normalization factor that keeps the norm of the wave function equal to unity. These formulae can be rewritten in the following form:

$$\psi_j(y) = \frac{E_j(y)\psi(y)}{\left[\int_{-\infty}^{+\infty} E_j(y)|\psi(y)|^2 dy \right]^{1/2}}, \quad (2.11)$$

where the set of functions $E_j(y)$ is defined by the rule

$$E_j(y) = \begin{cases} 1 & \text{if } y_j \leq y < y_{j+1} \\ 0 & \text{if } y < y_j \text{ or } y \geq y_{j+1} \end{cases}. \quad (2.12)$$

Notice that the functions $E_j(y)$ are the only things we need to know about the measuring apparatus in order to compute both the uncertainty in the results of the measurement and the back action of the measurement on the particle's wave function. The following two properties of the set of measuring-apparatus functions play a key role in the theory of measurement:

1. Completeness:

$$\sum_j E_j(y) \equiv 1 . \quad (2.13)$$

This completeness reflects the fact that the sum of the probabilities for all possible results of the measurement is unity. It is evident that this property must be satisfied by any measurement.

2. *Orthogonality:*

$$E_j(y)E_k(y) = \delta_{jk} E_j(y) . \qquad (2.14)$$

This property follows from the fact that the sieve's cells do not overlap each other, and it leads to the exact repeatability of the measurement's results: If, immediately after the first set of cells we shall put a second set, the particle that is registered as passing through cell j of the first set inevitably must pass through the corresponding cell of the second set.

The generalization of our sieve example to the measurement of an arbitrary observable q is straightforward. For concreteness, we presume that the spectrum of the observable q is continuous; if the spectrum is discrete, the integrations below must be replaced by summations. Suppose that the measuring device answers the question of which interval (q_j, q_{j+1}) the observable q lies in. The probability to be in interval j is equal to

$$w_j = \int_{q_j}^{q_{j+1}} <q|\hat{\rho}_{\text{init}}|q> dq = \text{Tr}(\hat{E}_j \hat{\rho}_{\text{init}}) , \qquad (2.15)$$

where $\hat{\rho}_{\text{init}}$ is the density operator describing the state of the measured object before the measurement, $|q>$ is the eigenstate of the measured observable, Tr denotes the trace (spur) of the indicated product of operators, and $\hat{E}_j$ is defined by

$$\hat{E}_j = \int_{q_j}^{q_{j+1}} |q><q| dq = \int_{-\infty}^{+\infty} |q>E_j(q)<q| dq , \text{ where}$$

$$E_j(q) = \begin{cases} 1 & \text{if } q_j \le q < q_{j+1} \\ 0 & \text{if } q < q_j \text{ or } q \ge q_{j+1} . \end{cases} \qquad (2.16)$$

This definition of the $\hat{E}_j$ guarantees completeness

$$\sum_j \hat{E}_j \equiv \hat{I} \qquad (2.17)$$

(where $\hat{I}$ is the identity operator) and orthogonality

$$\hat{E}_j \hat{E}_k = \delta_{jk} \hat{E}_j . \qquad (2.18)$$

These are generalizations of the completeness and orthogonality of equations (2.13) and (2.14).

The "reduced" state of the object after the measurement is given, in analogy with equations (2.8) and (2.11), by

$$\hat{\rho}_j = \frac{1}{w_j}\hat{E}_j\hat{\rho}_{\text{init}}\hat{E}_j \; . \tag{2.19}$$

The form of this reduced state guarantees repeatability of the measurement's results: after the measuring device has reported that q lies in the interval (q_j, q_{j+1}), an immediately subsequent measurement with an identical device must give the same result with probability unity. If, instead of measuring a second time which interval q lies in, one makes one's second measurement exact, the probability density for the result will be zero outside interval j, and inside that interval it will be given by the standard probability formula

$$w(q|j) = \frac{1}{w_j}\langle q|\hat{\rho}_{\text{init}}|q\rangle \; .$$

The set of operators (2.16) thus gives a complete characterization of the measuring apparatus. It permits one to compute both the probability distributions for the results of a measurement and the state of the object after the measurement. Because of the structure of the completeness relation (2.17), this set of operators is called a "decomposition of unity."

Measurements whose decomposition of unity satisfies $\hat{E}_j\hat{E}_k = \delta_{jk}\hat{E}_j$ [equation (2.18)] are called "orthogonal." As we have seen, repeated orthogonal measurements leave the state of an object unchanged. In this sense, orthogonal measurements are more nearly exact than any other type of measurement. In fact, they can be regarded as exact measurements of the operator

$$\hat{Q} = \sum_j q_j\hat{E}_j \; , \tag{2.20}$$

which has a degenerate set of eigenvalues (all states $|q\rangle$ with $q_j < q < q_{j+1}$ are eigenstates of $\hat{Q}$ with the eigenvalue q_j). This, in fact, is just the manner in which von Neumann introduced orthogonal measurements.[12] For orthogonal measurements of observables with the form (2.20), Equations (2.15) and (2.19) represent a mathematical reformulation of von Neumann's reduction postulate.

2.7* Nonorthogonal measurements

Let us return once more to measurements by the sieve apparatus of section 2.4 and Fig. 2.1, and ask about the changes in the particle's momentum P_y caused by the sieve's measurement of y. Suppose, for simplicity, that the initial state of the particle was $\psi(y) = \text{const}$, so P_y was well defined and equal to zero. Then after the measurement the wave function [according to equation (2.10)] will be

$$\psi_j(y) = \begin{cases} (y_{j+1}-y_j)^{-1/2} & \text{if } y_j \le y < y_{j+1} \\ 0 & \text{if } y < y_j \text{ or } y \ge y_{j+1} \end{cases}, \qquad (2.21)$$

The rms uncertainty in the particle's position in this state is

$$\Delta y = \frac{1}{2\sqrt{3}}(y_{j+1}-y_j) .$$

Thus, if this measurement were to produce the minimum perturbation of the momentum that is allowed by the uncertainty principle, then the momentum uncertainty after the measurement would be

$$\Delta P_y = \frac{\hbar}{2\Delta y} = \frac{\hbar\sqrt{3}}{y_{j+1}-y_j} .$$

However, as one can show by a simple calculation, the variance of the momentum in the state (2.21) is infinite. This peculiarity is a consequence of the discontinuity of the wave function (2.21), which in turn can be traced to the discontinuous character of the decomposition of unity (2.12); and it is a peculiarity characteristic of all orthogonal measurements.

Another peculiar property of orthogonal measurements is the repeatability of their results. This property is peculiar in that it contradicts the usual behavior of the state of an object when measurements are repeated: If a measurement does not perturb the observable being studied, then each successive measurement will usually give additional information about the value of the observable. The results of the successive measurements may differ from one another to within their errors, but our knowledge of the observable improves with each new measurement.

For orthogonal measurements this is not so. The first measurement gives full information about the observable, and the subsequent measurements can only deterministically repeat the result of the first.

These two peculiarities force one to conclude that orthogonal measurements are only some asymptotic, limiting case of realistic, approximate measurements. To describe realistic measurements we must move away from the orthogonal case in some manner.

The most general approach to the description of realistic quantum measurements is that of the *theory of optimal signal detection*.[22–25] This theory describes an arbitrary measurement in a manner analogous to the above treatment of orthogonal measurements: To each possible outcome $\tilde{q}$ of a measurement there corresponds an operator $\hat{\Pi}(\tilde{q})$ [generalization of $\hat{E}_j$]. Together these operators form a decomposition of unity analogous to

(2.17):

$$\int_{\{\tilde{q}\}} \hat{\Pi}(\tilde{q})d\tilde{q} = \hat{I} ,\qquad (2.22)$$

where $\{\tilde{q}\}$ is the set of all possible values of $\tilde{q}$. The $\tilde{q}$ can be a scalar, or a vector (if several observables are measured simultaneously), and it can have a continuous or a discrete spectrum. (For the discrete case one must replace the integrals by summations.) The probability distribution for the possible experimental outcomes is given by the same formula as in the orthogonal case:

$$w(\tilde{q}) = \text{Tr}[\hat{\Pi}(\tilde{q})\hat{\rho}_{\text{init}}] .\qquad (2.23)$$

The key difference of realistic measurements from orthogonal ones is the absence of any orthogonality constraint analogous to (2.18). The absence of this or any other constraints on the $\hat{\Pi}(\tilde{q})$, aside from the condition of completeness (2.22) [which is absolutely necessary to produce the correct normalization of the probabilities of the measurement outcomes], enables this formalism to describe all types of quantum measurements: exact measurements, approximate measurements, measurements of individual observables, and simultaneous measurements of several observables, including ones that do not commute with each other.

We shall not, in this book, delve into the full details of the general theory of decompositions of unity of the type (2.22). The full details are presented at the highest level of rigor in several monographs.[24,25] For us the most important feature of this general theory is the fact that it leaves the object's quantum state, after the measurement, undetermined. This is probably inevitable for so general a theory: to determine the object's final state in a realistic measurement, one must supply information about the details of the apparatus and measurement that probably are not fully incorporatable into the functions $\hat{\Pi}(\tilde{q})$. In view of this, we shall now specialize our discussion to approximate measurements of a single observable, and shall expand it to include more information about the measurement than is contained in the $\hat{\Pi}(\tilde{q})$.

The precision of an approximate measurement of a single observable depends on properties of the measuring device that we shall embody in a set of conditional probabilities $w(\tilde{q}|q)$. The function $w(\tilde{q}|q)$ tells us the probability that the measuring device will report a value $\tilde{q}$ when the observable is in an eigenstate with eigenvalue q. It is straightforward to show that, if the observable is in an initial state $\hat{\rho}_{\text{init}}$, then the *a priori* probability density for obtaining a value $\tilde{q}$ from the measurement has the form

$$w(\tilde{q}) = \int_{-\infty}^{+\infty} w(\tilde{q}|q)\rho(q)\,dq \,, \qquad (2.24)$$

where

$$\rho(q) = <q|\hat{\rho}_{\text{init}}|q> \qquad (2.25)$$

is the initial state's *a priori* probability distribution for the values of q. Equation (2.24) can be rewritten in a form analogous to equations (2.14) and (2.23):

$$w(\tilde{q}) = \text{Tr}[\hat{W}(\tilde{q})\hat{\rho}_{\text{init}}] \,, \qquad (2.26)$$

where the operators

$$\hat{W}(\tilde{q}) = \int_{-\infty}^{+\infty} |q>w(\tilde{q}|q)<q|\,dq \qquad (2.27)$$

make up the decomposition of unity that describes the measurement:

$$\int_{-\infty}^{+\infty} \hat{W}(\tilde{q})\,d\tilde{q} = \hat{I} \,. \qquad (2.28)$$

An important feature of this decomposition of unity is the fact that its elements commute with each other:

$$[\hat{W}(\tilde{q}), \hat{W}(\tilde{q}')] \equiv 0 \,. \qquad (2.29)$$

Because a necessary and sufficient condition for a set of operators to commute is the possibility to diagonalize them all simultaneously in a common representation, as is done in equation (2.27), any self-commuting decomposition of unity describes a measurement of a single observable (or of several commuting observables). As an example, the orthogonality condition (2.17) implies that the elements of any orthogonal decomposition of unity commute with each other, and thus orthogonal measurements are a subclass of the measurements we have been discussing. For them, the discontinuous functions (2.16) play the role of the conditional probabilities $w(\tilde{q}|q)$ that characterize the measuring apparatus.

2.8* Back action of the measuring device on the measured object

Continuing our discussion of approximate measurements of a single observable, we now turn our attention to the back action of the measuring device on the state of the measured object. By analogy with equation (2.19), we shall try to write the density operator of the final state of the object in the following form:

$$\hat{\rho}(\tilde{q}) = \frac{1}{w(\tilde{q})}\hat{\Omega}(\tilde{q})\hat{\rho}_{\text{init}}\hat{\Omega}^{\dagger}(\tilde{q}) \,, \qquad (2.30)$$

where $\hat{\Omega}(\tilde{q})$ is an operator, the form of which we shall try to determine. (We here have introduced the form (2.30) for the final state by a heuristic argument; we shall give a rigorous derivation of it in the next chapter.)

From the normalization condition $\mathrm{Tr}[\hat{\rho}(\tilde{q})] \equiv 1$ it follows that

$$\hat{\Omega}^{\dagger}(\tilde{q})\hat{\Omega}(\tilde{q}) = \hat{W}(\tilde{q}) \, . \tag{2.31}$$

Thus, $\hat{\Omega}(\tilde{q})$ can be represented in the form

$$\hat{\Omega}(\tilde{q}) = \hat{U}(\tilde{q})\hat{W}^{1/2}(\tilde{q}) \, , \tag{2.32}$$

where

$$\hat{W}^{1/2}(\tilde{q}) = \int_{-\infty}^{+\infty} |q{>}w^{1/2}(\tilde{q}\,|\,q){<}q\,|\,dq \tag{2.33}$$

is an operator that is uniquely determined by the measuring apparatus's conditional probabilities $w(\tilde{q}\,|\,q)$ and that commutes with the measured observable $\hat{q}$, and where $\hat{U}(\tilde{q})$ is some unitary operator.

Thus the transition from the initial state $\hat{\rho}_{\text{init}}$ to the final $\hat{\rho}(\tilde{q})$ (the state reduction) can be described by a two-step process: 1) the transformation $\hat{\rho}_{\text{init}} \rightarrow \hat{\rho}'(\tilde{q})$, where

$$\hat{\rho}'(\tilde{q}) = \frac{1}{w(\tilde{q})} W^{1/2}(\tilde{q})\hat{\rho}_{\text{init}}W^{1/2}(\tilde{q}) \, , \tag{2.34}$$

followed by 2) the unitary transformation

$$\hat{\rho}(\tilde{q}) = \hat{U}(\tilde{q})\hat{\rho}'(\tilde{q})\hat{U}^{\dagger}(\tilde{q}) \, . \tag{2.35}$$

Certainly, this theory does not assert that every real measurement truly consists of a two-step process. However, to each measurement one can associate a mathematical two-step state-reduction procedure that predicts the correct error of measurement and the correct back action on the object.

It is important to notice the following key feature of the first step, (2.34) of the state reduction. This step is completely determined by the *a priori* probability distributions $w(\tilde{q}\,|\,q)$, in other words, by the information that we get from the measurement. This step leaves the measured value of the observable unchanged: if it is applied to an ensemble of objects, then the probability distribution for the chosen observable after the measurements would be the same as before,

$$\int_{-\infty}^{+\infty} {<}q\,|\,\hat{\rho}'(\tilde{q})\,|\,q{>}w(\tilde{q})d\tilde{q} \equiv {<}q\,|\,\hat{\rho}_{\text{init}}\,|\,q{>} \, , \tag{2.36}$$

because the operator $\hat{W}^{1/2}(q)$ [equation (2.33)] commutes with the observable $\hat{q}$.

For so general an analysis as we are giving here, one cannot determine exactly the operator $\hat{U}(\tilde{q})$ that generates the second step of reduction. The precise form of that operator is determined by the concrete details of the measurement procedure. In general the measurement can perturb the quantity being measured, and this perturbation is one of the effects embodied in $\hat{U}(\tilde{q})$. However, because $\hat{U}(\tilde{q})$ is unitary, it cannot change the object's entropy, and correspondingly, this step of the reduction is not accompanied by the experimenter's extraction of any new information.

We have seen that, for a measurement of a single observable (or a group of commuting observables), and for any specific choice of the conditional probabilities $w(\tilde{q}\,|\,q)$, i.e. any specific level of precision in the apparatus, there exists a procedure of measurement that leaves the measured quantity unperturbed. Such a procedure is any one for which the operator $\hat{U}(\tilde{q})$ commutes with $\hat{q}$,

$$[\hat{U}(\tilde{q}), \hat{q}] = 0 . \tag{2.37}$$

The property of leaving the measured quantity unperturbed is embodied in equation (2.36), with the semi-final density operator, $\hat{\rho}'(\tilde{q})$, replaced by the final one, $\hat{\rho}(\tilde{q})$. Such measurements have been given the name quantum nondemolition (QND) measurements; for details see section 4.2.

Observables that do not commute with the measured quantity $\hat{q}$ must, according to the uncertainty relations, be perturbed by the measurement. The strength of this perturbation, as one can see from equation (2.31), is uniquely determined by the form of the conditional probabilities $w(\tilde{q}\,|\,q)$, i.e. by the information that we can obtain in the process of the measurement.

III   Indirect measurements

3.1 The two main types of quantum measurements

von Neumann's postulate of the reduction of the wave function answers the question of *what* happens to the quantum object during the measurement. However, the reduction postulate does not answer the question of *how* the measuring device must be designed to realize the desired measurement. To answer this second question, one must understand the connection between the measuring device as an ordinary physical system, and the nature of the desired measurement (the quantity to be measured, the desired precision, and the choice of how the measurement is to perturb the quantum object).[*]

The Schrödinger equation cannot tell us the connection between the design of the measuring device and the nature of the measurement, because the Schrödinger equation neither governs nor describes the process of measurement. More specifically: The evolution of the wave function as described by the Schrödinger equation has two key features: it is a reversible evolution (from the final state in principle one can always evolve back to the initial one), and it is a deterministic evolution (the final state is determined uniquely by the initial one). By contrast, the reduction of the wave function in a measurement is irreversible (once the

---

[*]The authors follow the traditional (von Neumann) interpretation of quantum mechanics. For a review of other interpretations see Ref. 73.

measurement is finished and the information has been extracted from it, it is impossible to return to the pre-measurement state), and it is nondeterministic (one cannot predict, before the measurement, the post-measurement state of the measured object). The only exception to these properties of the reduction is a perfect quantum nondemolition (QND) measurement (see the next chapter), in a case where the object is initially in an eigenstate of the measured observable. (Such a measurement is useful only if the experimenter does not know in advance the state's eigenvalue—for example, the experimenter might prepare the object in a known eigenstate with known eigenvalue, then an unknown classical force might act, changing the eigenvalue in an unknown way but leaving the object in an eigenstate of the observable, and then the experimenter might perform a QND experiment to determine the new eigenvalue and from it learn about the classical force. von Neumann's reduction procedure in this special case leaves the object's wave function unchanged.)

The impossibility of describing the quantum measurement process within the framework of the standard evolutionary apparatus of quantum mechanics can be traced to the fact that the receiver of the measurement's information is a macroscopic observer. The question of whether quantum mechanics can describe fully the macroscopic world is an open one, and is widely discussed by physicists and philosophers. One thing, however, is evident: The observer cannot even in principle write down a wave function for himself. To do so, it would be necessary to define the initial states of all the elementary particles from which the observer is made. However, the amount of information that the observer is able to comprehend is determined by the number of neurons in his brain ($\sim 10^{10}$). This number is far smaller than the number of particles that make up his brain and body (after all, each neuron itself is made out of a large number of elementary particles), so he cannot possibly comprehend the content of a wave function that describes himself.

On the other hand, it would be incorrect to suppose that to analyze the procedure of measurement, one must always include explicitly the observer. In general, between the observer and the quantum object there is a so-called classical measuring device—an inanimate system with many degrees of freedom ($\sim 10^{23} \gg 10^{10}$), which under the action of the measured quantum object irreversibly changes its own state in a manner that the observer can directly comprehend. Examples of such classical measuring devices are the photoemulsion of a photographic plate, the supersaturated steam in a Wilson cloud chamber, and the active medium in a quantum amplifier. There is even an example in which the device was the human eye: Pavel Cerenkov's discovery of Cerenkov radiation in which he detected individual photons directly with his eyes.

Thermodynamics tells us that all these complicated, classical measuring devices are irreversible, and the detailed motions of their constituent particles are unpredictable.

Quantum measurements in which the measured object interacts directly with the classical measuring device are called "direct measurements." A typical example of a direct measurement is the measurement of the position of a microparticle by the path it produces in a photoemulsion. In a direct measurement, there typically is considerable randomness in the interaction of the object with the many degrees of freedom of the classical measuring device (e.g. the random motion of the microparticle in the photoemulsion, which resembles Brownian motion). As a result, the measuring device perturbs the object far more strongly than the minimum perturbation required by the uncertainty relations. Moreover, the device usually perturbs not only the degree of freedom that it measures, but also other degrees of freedom. For example, the photoemulsion perturbs not only the components of the microparticle's momentum parallel to the photographic plate (the momenta whose corresponding coordinates are being measured), but also the perpendicular component of the momentum. Such perturbations are inevitable because many different degrees of freedom of the photoemulsion participate in the random extraction of energy from the microparticle. The energy absorbed by each photoemulsion degree of freedom is rather large: typically several electron volts, and the same is true for the Wilson cloud chamber and other typical direct measurement devices. When the microparticle is an optical photon (which has a total energy of only a few electron volts), this interaction causes the photon to be completely absorbed.

Much better results can be obtained by using a second type of quantum measurement: an "indirect measurement" (a term coined by Leonid Mandelstam[26]). An indirect measurement is a two-step process. In the first step the object interacts with a quantum system that has been prepared in advance in some special initial quantum state. This quantum system is called the "quantum probe." There is no reduction of any state in this first step of the measurement, since it is an ordinary interaction between two quantum systems, governed by the Schrödinger equation. The second step is a direct measurement of some chosen observable of the quantum probe. The state of the probe is reduced in this second step, and since the first step produced a correlation between the state of the probe and the state of the quantum object, there is enforced a corresponding reduction of the object's state. Correspondingly, in the end one can neither counteract the back action that the probe has had on the quantum object, nor even know precisely what that back action was (except by a subsequent measurement). Thus, the measurement has produced an irreversible and indeterminate change in the quantum object.

When analyzing an indirect measurement, one can typically place the boundary between the quantum probe and the rest of the measuring system (the so-called "quantum/classical cut") at several different locations. Where one places it will not influence one's final conclusion about the accuracy of the experiment or the reduction of the object's state.

To achieve high precision in an indirect measurement, it is a good strategy to choose an experimental design that satisfies two conditions: *First*, the second step of the measurement (interaction of the quantum probe with the remaining, classical part of the measuring device) should not begin until the first step (interaction of object with probe) is complete. This condition isolates the measured object from the "killing-of-the-quantum-state" influence of the classical device. *Second*, the second step should not contribute significantly to the total error of the measurement. An example is the Heisenberg microscope, where the probe is the photon, the first step is the photon scattering off the measured object, passing through the lens, and flying toward the photographic plate; and the second step is the photon hitting the photographic plate, there being destroyed and making a silver embryo, and the development of the plate to grow a silver grain from the embryo. In this example the experiment should be so designed that the finite size of the silver grain contributes a negligible amount to the error in the inferred position of the object. The error should be dominated by the diffraction limit of the lens [Eq. (1.11)].

When these two conditions are satisfied, then the only source of error in the measurement, and the only perturbation of the object, will be those due to the internal uncertainties of the quantum probe. In the ideal case, these will be the quantum uncertainties of the probe's initial state, as enforced by the Heisenberg uncertainty principle. In a real physical experiment, however, additional errors may arise, e.g. from dissipation in the quantum probe. In any case, if the above two conditions are satisfied, the magnitudes of the error and the perturbation can be inferred from an analysis solely of the *first step* of the measurement, i.e., from a solution of the Schrödinger equation for the mutual evolution of the quantum probe and the measured object. All we need to know about the second step is what observable of the quantum probe it registers.

3.2 An electron as the quantum probe

In this section we shall illustrate the above concepts by a simple device for indirect measurements: one that uses an electron as the quantum probe.

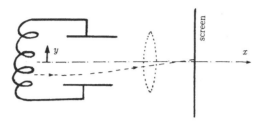

Fig. 3.1 Measurement of the charge on a capacitor using an electron probe.

   The goal of an experiment with this device is to measure the electric charge on the plates of a flat capacitor, which is a part of the quantum object (e.g., an **LC** circuit). To make this measurement, a charged probe particle (for example, an electron) is launched through the capacitor as shown in Fig. 3.1. If the time τ spent traversing the capacitor is sufficiently small, the charge q on the capacitor will remain essentially constant during the traversal, and the change in the y-component of the electron's momentum will be proportional to q:

$$P_y^{\text{final}} = P_y^{\text{init}} + kqy, \qquad (3.1)$$

where

$$k = \frac{e\tau}{Cd}$$

is the coupling constant of the probe to the capacitor (with e the electron charge, C the capacitance, and d the distance between the plates). By measuring P_y^{final}, one can determine q with the precision

$$\Delta q = \frac{1}{k}[(\Delta P)_1{}^2 + (\Delta P)_2{}^2]^{1/2} \, ,$$

where $(\Delta P)_1$ is the initial uncertainty in the electron's momentum P_y^{init}, and $(\Delta P)_2$ is the error in the measurement of P_y^{final}.

   To measure the momentum P_y^{final} after the electron traverses the capacitor, the experimenter sends the electron through an electronic lens and onto a photographic plate situated at the lens's focal plane. The coordinate y' of the point a at which the electron hits the plate is (cf. Fig. 3.1)

$$y' = f\frac{P_y^{\text{final}}}{P_x} \, ,$$

where f is the lens's focal length and P_x is the x component of the electron's momentum. Correspondingly,

$$(\Delta P)_2 = \frac{P_x}{f} \cdot \Delta y' \,,$$

where $\Delta y'$ is the size of the spot on the plate. By increasing the focal length, one can ensure that $(\Delta P)_2 \ll (\Delta P)_1$, so the error in the measurement of the capacitor's charge is determined solely by the initial state of the electron,

$$\Delta q \simeq \frac{(\Delta P)_1}{k} \,. \tag{3.2}$$

During the electron's traversal through the capacitor, it induces on the capacitor plates an electromotive force proportional to the electron's y coordinate. One can show that this force perturbs the generalized momentum Φ of the measured object [if the object is an LC circuit, then Φ is the magnetic flux in the inductor] and the uncertainty in the strength of this perturbation is

$$\Delta \Phi = k \, \Delta y \,, \tag{3.3}$$

where Δy is the uncertainty in y. The product of the measurement error (3.2) and the perturbation (3.3), as one can easily see, is in precise agreement with the Heisenberg uncertainty relation:

$$\Delta q \cdot \Delta \Phi = (\Delta P)_1 \cdot \Delta y \geq \frac{\hbar}{2} \,. \tag{3.4}$$

Thus, by analyzing the interaction of the probe electron with the measured object (first step of measurement), we can calculate the principal characteristics of the measuring procedure—the measurement accuracy and the magnitude of the back action.

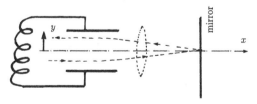

Fig. 3.2 Electron-probe experiment to measure the charge on a capacitor (Fig. 3.1), with the photographic plate replaced by a mirror.

If, however, the actual experiment is limited to the first step, it loses its key properties of irreversibility and indeterminism. For example, if we replace the photographic plate by a mirror for electrons (as shown in Fig. 3.2), then the electron after reflection will fly through the capacitor a second time and can completely undo the perturbation that it gave to the measured object. In this case, the electron itself will also return to its initial (but symmetrically reversed) state, and all the information about the capacitor's charge q, which was stored in the electron's state during the first traversal, will be removed.

If, instead of reflecting the electron off a mirror (Fig. 3.2) or using a lens and photographic plate to infer its momentum (Fig. 3.1), we measure the electron's transverse position y immediately after it traverses the capacitor, then from the result we can infer the strength of the electron's back action on the measured object's generalized momentum Φ. In this case there will be a unique, deterministic coupling between the initial and final states of the object, but only at the expense of the electron having lost all information about the value of the capacitor's charge q (because the measurement of the electron's y coordinate strongly perturbs the momentum P_y which was carrying the information about q).

3.3* Electron probe — detailed analysis

The purpose of this section is to illustrate, using the simple example of the last section, a method by which one can analyze indirect measurements in detail.

To avoid complicating the analysis by inessential details, we shall presume that the time τ for the electron's traversal through the capacitor is so small that one can neglect the free evolution of both the electron and the measured object during the traversal. Then during the traversal the complete Hamiltonian for the system (electron plus object) reduces to just the interaction part of the Hamiltonian,

$$\hat{H} = -\frac{e\hat{q}\hat{y}}{Cd} . \tag{3.5}$$

If the electron initially is in the pure state $|\psi\rangle$ and the object is in the mixed state $\hat{\rho}_{\mathrm{init}}$, then after the traversal their joint state will be described by the density operator

$$e^{-i\hat{H}\tau/\hbar}|\psi\rangle\hat{\rho}_{\mathrm{init}}\langle\psi|e^{i\hat{H}\tau/\hbar} . \tag{3.6}$$

After the electron leaves the capacitor, its momentum is measured. According to the description in the last section, this measurement can be regarded as exact. According to the reduction postulate, the probability that this exact measurement will give a value P for the momentum P_y is

$$W(P) = \mathrm{Tr}\left[|P\rangle\langle P|e^{-i\hat{H}\tau/\hbar}|\psi\rangle\hat{\rho}_{\mathrm{init}}\langle\psi|e^{+i\hat{H}\tau/\hbar}\right] . \tag{3.7}$$

Using the property

$$e^{i\xi\hat{y}/\hbar}|P> = |P+\xi>$$ (3.8)

of the shift operator, equation (3.7) can be rewritten in the form

$$W(P) = \int_{-\infty}^{+\infty} <q|\hat{\rho}_{\text{init}}|q> w_P(P-eq\tau/Cd)dq \; ,$$ (3.9)

where $w_P(P)$ is the probability distribution for the electron's momentum P_y in the initial state. Using the formula

$$\tilde{q} = \frac{Cd}{e\tau}(P-\bar{P})$$ (3.10)

for the inferred value $\tilde{q}$ of the capacitor's charge in terms of the measured value P of the electron's momentum (where $\bar{P}$ is the expectation value of the momentum in the initial state), we obtain the following form for the probability distribution of $\tilde{q}$:

$$w(\tilde{q}) = \int_{-\infty}^{+\infty} <q|\hat{\rho}_{\text{init}}|q> \frac{e\tau}{Cd} w_P\left[\frac{e\tau}{Cd}(\tilde{q}-q)+\bar{P}\right] dq \; .$$ (3.11)

By comparing equations (3.11) and (2.24) one sees that

$$\frac{e\tau}{Cd} w_P\left[\frac{e\tau}{Cd}(\tilde{q}-q)+\bar{P}\right]$$ (3.12)

is the conditional probability $w(\tilde{q}|q)$ which, via formula (2.27), determines the decomposition of unity for this measurement. The variance of this conditional probability distribution describes the spread that the results of the measurement would have if q were precisely defined before the measurement. In other words, the square root of this variance is the rms measurement error Δq. By computing the variance we see that

$$\Delta q = \frac{Cd}{e\tau}\Delta P \; ,$$ (3.13)

where $(\Delta P)^2$ is the variance of the initial probability distribution w_P for the electron's momentum.

After the measurement, the object's state will have been reduced to

$$\hat{\rho}_{\text{final}} = \frac{1}{W(P)} <P|e^{-i\hat{\Pi}\tau/\hbar}|\psi> \hat{\rho}_{\text{init}} <\psi|e^{+i\hat{\Pi}\tau/\hbar}|P> \; .$$ (3.14)

By using equations (3.8) and (3.10) once again, one can bring this density operator into the form (2.30) with

$$\hat{\Omega}(\tilde{q}) = \int_{-\infty}^{+\infty} |q> \sqrt{\frac{e\tau}{Cd}} \psi_P\left[\frac{e\tau}{Cd}(\tilde{q}-q)+\bar{P}\right] <q|dq \; ,$$ (3.15)

where $\psi_P(P) = \langle P | \psi \rangle$ is the electron's initial wave function in the momentum representation.

This operator $\hat{\Omega}(\tilde{q})$, as should be evident from (3.15), commutes with the observable $\hat{q}$ being measured. Thus, the charge q is unchanged by the measurement. However, the generalized momentum Φ, which is canonically conjugate to q, *is* perturbed. The perturbation can be computed by presuming that the initial state of the object has a definite value Φ_o of Φ, and by then computing the variance of Φ after the measurement:

Substituting $\hat{\rho}_{\text{init}} = | \Phi_o \rangle\langle \Phi_o |$ into equation (3.6) and using the relation

$$e^{i\xi\hat{q}/\hbar}| \Phi \rangle = | \Phi + \xi \rangle$$

[analogous to equation (3.8)], we conclude that, at the end of the interaction between the electron and the object their joint wave function has the form

$$| \psi_{\text{final}} \rangle = \int_{-\infty}^{+\infty} \psi_y(y) | y | \Phi_o + ey\,\tau/Cd \rangle dy \ . \tag{3.16}$$

Here $\psi_y(y) = \langle y | \psi \rangle$ is the initial wave function of the electron in the coordinate representation. To the wave function (3.16) corresponds the following probability distribution for Φ:

$$w_{\text{final}}(\Phi) = \int_{-\infty}^{+\infty} | \langle \Phi | y | \psi_{\text{final}} \rangle |^2 dy = \frac{Cd}{e\,\tau} w_y \left[\frac{Cd}{e\,\tau}(\Phi - \Phi_o) \right] \ . \tag{3.17}$$

Here $w_y(y) = | \psi_y(y) |^2$ is the probability distribution for y in the initial state. This $w_{\text{final}}(\Phi)$ implies that the rms perturbation of Φ is

$$\Delta\Phi = \frac{e\,\tau}{Cd}\Delta y \ ,$$

where Δy is the initial uncertainty in y.

3.4[*]   Formal description of an indirect measurement

This section presents a general theory of indirect measurements. This theory will be both a straightforward generalization of the electron probe example of the last section, and a specialization of section 2.7's formal theory of approximate, nonorthogonal measurements. Some features of the approximate, nonorthogonal measurement theory, which may have seemed to be "pulled out of the air" in section 2.7, will be made clearer and more compelling by this theory of indirect measurements.

We shall restrict attention to indirect measurements which satisfy the two conditions discussed near the end of section 3.1: the first step ends before the second begins, and the second step contributes negligibly to the error of measurement.

Before the measurement, the quantum probe is prepared in a carefully selected initial state, which we shall describe by the density operator $\hat{\rho}_{probe}$, and the measured object is in an initial state described by the density operator $\hat{\rho}_{init}$. The first step of the measurement is the interaction between probe and object. This interaction brings the probe and object into the joint, correlated quantum state

$$\hat{U}\hat{\rho}_{init}\hat{\rho}_{probe}\hat{U}^{\dagger} \; , \tag{3.18}$$

where $\hat{U}$ is the unitary operator that evolves them from the beginning of their interaction to the end. The corresponding state of the probe alone, after the interaction, is

$$\mathrm{Tr}_{obj}(\hat{U}\hat{\rho}_{init}\hat{\rho}_{probe}\hat{U}^{\dagger}) \; , \tag{3.19}$$

where Tr_{obj} denotes the trace over the object's Hilbert space.

The second step is a direct measurement of some carefully chosen observable P of the probe. Since this measurement contributes negligibly to the experiment's overall error, we can idealize it as arbitrarily accurate. From the design of the experiment, the experimenter infers a one-to-one correspondence between the outcome P of this precise measurement on the probe, and the value $\tilde{q}$ of the object's observable, which is the final result of the experiment. Because of this one-to-one correspondence, we can henceforth use $\tilde{q}$ as a surrogate for P, i.e. we can regard it not only as the inferred value of the object's observable, but also as the result of the precise, step-two measurement of the probe. The eigenstate of the probe's precisely measured observable with eigenvalue $\tilde{q}$ (surrogate for P) we shall denote by $|\tilde{q}>$, and the operators that make up the decomposition of unity for the precise step-two measurement of the probe are $|\tilde{q}><\tilde{q}|$, with $\tilde{q}$ allowed to take on all possible values that the experiment could have produced. By inserting into equation (2.23) this decomposition of unity and the density operator (3.19) for the probe just before the precise step-two measurement, we obtain the following probability distribution for the result of the measurement:

$$w(\tilde{q}) = \mathrm{Tr}_{probe}\left[|\tilde{q}><\tilde{q}|\,\mathrm{Tr}_{obj}(\hat{U}\hat{\rho}_{init}\hat{\rho}_{probe}\hat{U}^{\dagger})\right] \; , \tag{3.20}$$

where Tr_{probe} is the trace over the probe's Hilbert space. By a simple manipulation this formula can be rewritten as

$$w(\tilde{q}) = \mathrm{Tr}_{obj}[\hat{\Pi}(\tilde{q})\hat{\rho}_{init}] \; , \tag{3.21}$$

where

$$\hat{\Pi}(\tilde{q}) = \text{Tr}_{\text{probe}}\left[\hat{\boldsymbol{U}}^{\dagger} | \tilde{q} \rangle \langle \tilde{q} | \hat{\boldsymbol{U}} \hat{\rho}_{\text{probe}} \right] . \tag{3.22}$$

In equations (3.21) and (3.22) we are invited to regard $\tilde{q}$ as the outcome of the entire two-step measurement: the inferred value of the object's observable. We can then make contact with the general theory of approximate, nonorthogonal measurements (sections 2.7 and 2.8). By comparing equations (3.21) and (2.26), we see that the operators $\hat{\Pi}(\tilde{q})$, with $\tilde{q}$ ranging over all possible experimental outcomes, make up the decomposition of unity for the entire two-step measurement. Equation (3.22) expresses this $\hat{\Pi}(\tilde{q})$ in terms of the probe's initial state $\hat{\rho}_{\text{probe}}$, the final eigenstate $| \tilde{q} \rangle$ of the probe corresponding to the experimental outcome, and the evolution operator $\hat{\boldsymbol{U}}$ for the first step of the measurement.

The back action of the entire two-step measurement on the measured object is embodied in the object's final state. In section 2.8 it was asserted that this final state should have the form (2.30), but the proof was left until now. We can derive that form in the following way:

The above considerations imply that the density operator of the object's final state, after step two has produced the precise outcome $\tilde{q}$, is

$$\hat{\rho}(\tilde{q}) = \frac{1}{w(\tilde{q})} \langle \tilde{q} | \hat{\boldsymbol{U}} \hat{\rho}_{\text{init}} \hat{\rho}_{\text{probe}} \hat{\boldsymbol{U}}^{\dagger} | \tilde{q} \rangle . \tag{3.23}$$

As an aid in bringing this into the form (2.30), we shall express the probe's initial density operator in the form

$$\hat{\rho}_{\text{probe}} = \sum_{j} | \psi_j \rangle w_j \langle \psi_j | , \tag{3.24}$$

where w_j is the initial classical probability for the probe to be in the pure quantum state $| \psi_j \rangle$. By substituting this expression into equation (3.23) we obtain

$$\hat{\rho}(\tilde{q}) = \frac{1}{w(\tilde{q})} \sum_{j} w_j \hat{\Omega}_j(\tilde{q}) \hat{\rho}_{\text{init}} \hat{\Omega}_j^{\dagger}(\tilde{q}) , \tag{3.25}$$

where

$$\Omega_j(\tilde{q}) = \langle \tilde{q} | \hat{\boldsymbol{U}} | \psi_j \rangle .$$

The object's final density operator (3.25) is a mixture of states of the type (2.30), with the weighting factors w_j. As a result, the operator's final state has additional uncertainties, not taken into account in our formal theory of approximate, nonorthogonal measurements—uncertainties produced by the fact that here we allowed the probe to begin in a mixed state, but there it began in a pure one. Thus, equation (3.25) takes account not only of back-action uncertainties that are of an inevitable quantum origin, but also back-action uncertainties due to such effects as a finite initial temperature of the probe. If there are no such additional uncertainties, i.e. if the probe begins in a pure quantum state

$$\hat{\rho}_{probe} = |\psi><\psi| , \qquad (3.26)$$

then the object's final density operator has the form (2.30), and the operator $\hat{\Omega}(\tilde{q})$ is the amplitude for the probe to have evolved, during the first step of the measurement, from $|\psi>$ to $|\tilde{q}>$:

$$\hat{\Omega}(\tilde{q}) = <\tilde{q}|\hat{U}|\psi> . \qquad (3.27)$$

Thus, we have derived equation (2.30) for the object's final density operator, we have learned the conditions under which it is valid (the absence of additional classical uncertainties in the initial state of the probe), and we have learned the connection between the structure of the probe and the form of the operator $\hat{\Omega}(\tilde{q})$, which via equation (2.30) governs the measurement-induced reduction of the object's state.

IV Quantum nondemolition measurements

4.1 The standard quantum limit for the energy of an oscillator

In the 1970s, in connection with efforts to construct detectors for gravitational waves, it became necessary to invent methods for measuring macroscopic observables at levels of precision approaching and exceeding the standard quantum limits (cf. section 1.4). However, theoretical analyses of typical, traditional schemes of measurement showed that their precisions can never exceed the standard quantum limit, even in principle. The solution to this dilemma, it was recognized, was to use a nontraditional class of measurement schemes, carefully crafted to overcome the standard quantum limits. For these schemes was coined the term "quantum nondemolition measurements".[16, 27]

The task of measuring the energy in an electromagnetic resonator provides us, in this section, with an example of a traditional scheme of measurement and its inability to overcome the oscillator's standard quantum limit. At and below microwave frequencies, the traditional method of measuring a resonator's energy is to measure the amplitude of oscillation, by sending a signal from the resonator through a linear amplifier, and to then compute the energy from the amplitude. Now, it is evident that along with the amplitude, one can also extract from the amplifier's output the resonator's phase. Recall that the energy and phase are canonically conjugate observables (the values of the amplitude and the phase completely determine the classical state of the resonator, just as do a generalized coordinate and momentum.) However, as we have seen earlier, the precision with which one can know, simultaneously, the values of two canonically conjugate observables is limited by the uncertainty relation, which in this case reads

$$\Delta E \Delta \phi \geq \frac{\hbar \omega}{2} ,$$

where ω is the frequency of the resonator.

The "mechanism" by which the measurement process enforces this limit is similar to that in the Heisenberg microscope example of section 1.4: When the amplifier (a macroscopic, classical device) extracts information about the phase, it perturbs the resonator's energy, and conversely, when it extracts information about the energy, it perturbs the phase. The more accurately the experiment monitors the resonator (the stronger is the coupling between the resonator and amplifier), the larger is the back-action perturbation. There is an optimal coupling strength at which the accuracy of the monitoring and the strength of the perturbation are balanced, and the resulting total error in the measurement is minimized. The following analysis shows that the resulting minimum total error in the measured value of the energy is equal to the standard quantum limit, $\sqrt{\hbar \omega E}$ [equation (1.25)].

The generalized coordinate q of the chosen mode of the resonator (the charge q on the capacitor if the resonator is an *LC* circuit) oscillates in time as

$$q(t) = q_1 \cos\omega t + q_2 \sin\omega t . \qquad (4.1)$$

The quantities q_1 and q_2 are usually called the mode's quadrature amplitudes. They are connected to the initial values of the generalized coordinate q and its corresponding generalized momentum Φ by

$$q_1 = q(0) , \qquad q_2 = \frac{1}{\rho}\Phi(0) ,$$

where ρ is the so-called wave impedance of the resonator. (For an *LC* circuit, Φ is the magnetic flux in the inductor, and ρ is the square root of the ratio of the inductance to the capacitance, $\rho = \sqrt{L/C}$.) From these relations and the standard uncertainty relation for q and Φ, we infer that the quadrature amplitudes must satisfy the uncertainty relation

$$\Delta q_1 \cdot \Delta q_2 \geq \frac{\hbar}{2\rho} . \qquad (4.2)$$

The monitoring of $q(t)$ with a precision that is independent of time is clearly equivalent to measuring the quadrature amplitudes with equal precisions

$$\Delta q_1 = \Delta q_2 . \qquad (4.3)$$

By combining this with the uncertainty relation (4.2), we obtain the following fundamental limits on the accuracy with which this measurement scheme can determine the generalized coordinate and the quadrature amplitudes:

$$\Delta q_{SQL} = \Delta q_{1\,SQL} = \Delta q_{2\,SQL} \geq \sqrt{\frac{\hbar}{2\rho}} . \qquad (4.4)$$

These are the standard quantum limits for q, q_1, and q_2.

From a continuous monitoring of the coordinate q with a precision Δq we can learn the amplitude $A = \sqrt{q_1^2 + q_2^2}$ of the oscillations with this same precision:

$$\Delta A = \Delta q \geq \sqrt{\frac{\hbar}{2\rho}} . \qquad (4.5)$$

The energy is connected to this amplitude by the formula

$$E = \frac{\rho \omega A^2}{2} . \qquad (4.6)$$

If the error in the amplitude is small enough, $\Delta A \ll A$, then this formula implies an error in the energy given by

$$\Delta E \simeq \rho \omega A \, \Delta A . \qquad (4.7)$$

By a simple transformation of this formula and taking account of the limit (4.5) on ΔA, we find the following form for the ultimate limit on the accuracy with which this measuring scheme can measure the oscillator's energy (the standard quantum limit):

$$\Delta E_{SQL} \simeq \sqrt{\hbar \omega E} . \qquad (4.8)$$

Note that this standard quantum limit can be rewritten in the form

$$\Delta E_{SQL} \simeq \hbar \omega \sqrt{n} , \qquad (4.9)$$

where n is the mean number of quanta in the oscillator, i.e. $E = \hbar \omega (n + \frac{1}{2})$. Note that the condition $\Delta A \ll A$ under which these formulae are valid translates into $n \gg 1$, i.e. the mean number of quanta in the oscillator must be large. However, the standard quantum limits (4.4) on the quadrature amplitudes are precise, independently of the oscillator's level of excitation.

Let us emphasize once more that this standard quantum limit cannot be overcome by improving the quality of the amplifier, decreasing the temperature, or any other such strategy. This limit is a fundamental one for any measuring device that monitors the full sinusoidal oscillations of the resonator (i.e. any device that gives information about both the amplitude and the phase, i.e. that measures the quadrature amplitudes with equal precisions).

4.2 How can one overcome the standard quantum limit?

In order to measure the energy of a resonator with an accuracy higher than the standard quantum limit, a measuring device must avoid "seeing" the resonator's phase; it must respond only to the energy. During the measurement the device's back action will perturb the phase, but the device's output signal will not record any consequences of the perturbation. Such a device can completely avoid perturbing the resonator's energy because it avoids collecting any information at all about the phase.

A well known example of a device that measures electromagnetic energy without responding to the phase is a photon counter. (The fact that photon counters are typically used at much higher electromagnetic frequencies than linear amplifiers is irrelevant to our discussion, since we here are concerned only with matters of principle.) To measure the energy in an electromagnetic resonator, one can couple a photon counter to it and wait long enough to be sure that all the photons have been absorbed. If the quantum efficiency of the counter is close enough to unity, then with high probability it will count exactly the number of quanta in the resonator, and from that number and the energy $\hbar\omega$ of one photon, one can calculate the resonator's initial energy. The precision of such a measurement is not constrained by the standard quantum limit.

The evident disadvantage of this method of measurement is that all the energy in the resonator is absorbed; i.e., the back action on the resonator is enormously larger than when a linear amplifier is used. The cause of this perturbation is completely different from that in the linear amplifier case: instead of being due to the uncertainty principle, it is due to the fact that the measurement is a direct one: The photon counter's photocathode is a system with a very large number of degrees of freedom and it thus has thermodynamically unpredictable behavior and completely "demolishes" the quantum state of the resonator.

The principle underlying a method of measuring electromagnetic energy more accurately than the standard quantum limit and avoiding photon absorption has been known for approximately 100 years: Measure the electromagnetic pressure associated with the energy by an experiment like that of Peotr N. Lebedev, and from that pressure, infer the energy. Figure 4.1 shows such a scheme. One of the walls of the resonator is flexible or movable (like a plunger or piston in a cylinder). From the force on this wall one infers the electromagnetic pressure, and thence the

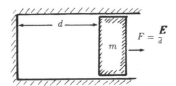

Fig. 4.1 The measurement of the electromagnetic energy in a resonator by its ponderomotive pressure on a movable wall.

resonator's energy. If the wall's inertia is large enough, then the phase of the electromagnetic oscillations will not influence the wall's motion. The motion will be slow compared to the electromagnetic frequency, i.e. the motion will be adiabatic; and it is known that the number of quanta in the resonator is left unaffected when the cavity is adiabatically deformed.

The ponderomotive (i.e. electromagnetic pressure) force acting on the resonator's wall is

$$F = \frac{E}{d} , \tag{4.10}$$

where E is the resonator's energy and d is a length of the order of the resonator's size that depends on which mode of the resonator is excited. The force F during a time τ (the "measurement time") changes the wall's momentum by an amount

$$\delta P = \frac{E\tau}{d} . \tag{4.11}$$

Thus, the more precisely the initial momentum is defined, the more precise will be the measured energy:

$$\Delta E_{\text{measure}} = \frac{d}{\tau}\Delta P , \tag{4.12}$$

where ΔP is the initial uncertainty of the momentum.

On the other hand, according to the uncertainty principle, the smaller is ΔP, the larger is the initial uncertainty Δx in the wall's position. This position uncertainty produces a corresponding uncertainty in the electromagnetic frequency during the measurement

$$\Delta \omega = \omega \frac{\Delta x}{d} , \tag{4.13}$$

and this in turn produces a random change in the resonator's phase during the measurement

$$\Delta \phi_{\text{perturb}} = \Delta \omega \cdot \tau = \omega \tau \frac{\Delta x}{d} . \tag{4.14}$$

From equations (4.10) and (4.14), and the uncertainty relation $\Delta x \Delta P \geq \hbar/2$, we obtain

$$\Delta E_{\text{measure}} \cdot \Delta \phi_{\text{perturb}} \geq \frac{\hbar \omega}{2} . \tag{4.15}$$

A more detailed analysis of this scheme of measurement will be presented in the next section. There it will be shown that the minimum error in the measured energy $\Delta E_{\text{measure}}$ is of order $\hbar/\tau$. Thus, the error of the measurement is not constrained by the standard quantum limit; by taking long enough for the measurement, one can obtain any desired sensitivity.

This scheme of measurement has all the properties of an ideal quantum measurement, as formulated at the end of section 2.4:

1. There is no fundamental constraint, analogous to the standard quantum limit, on the measurement precision.

2. There need not be any perturbation of the measured observable itself.

3. The quantity canonically conjugate to the measured observable (in this case the resonator's phase) is perturbed in accord with the uncertainty principle.

Measurements of this type have been given the name "quantum nondemolition" (QND) measurements. This name emphasizes property 2: the fact that the measured quantity is left unperturbed.

From the above discussion the general structure of a quantum nondemolition measurement should be evident: First, the object's only contact with the measuring device should be an interaction with a quantum probe (in the above case, the wall of the resonator), which has been prepared in a special quantum state (in the above case, a state of well defined momentum). Second, the interaction between the probe and the object must be organized in such a way that the probe is influenced only by one observable, or a set of observables that are not affected by the probe's back action on the object. This second condition, reexpressed in rigorous quantum language, says that the object observables that influence the probe must all commute with each other.

The ponderomotive example illustrates the principal difficulty that an experimenter must confront in trying to realize a QND measurement: It is necessary in the example to register the pressure produced by a few quanta, and this is an extremely small pressure, even at optical frequencies. More generally, in any QND experiment, the signal that the experimenter must deal with is extremely weak. Its weakness in an inevitable consequence of the character of the interaction between the object and the probe.

One can easily see that some observables cannot be measured in a QND way. An example is the position of a free mass (e.g. a free particle). Although the position need not be perturbed directly by the measurement, the momentum will be perturbed, and that perturbation will cause the mass's wave function to spread, thereby making the position uncertain if the measurement lasts for any finite time. Generalizing this, we see that QND measurements can be made only on observables that are conserved during the object's free evolution, in other words, only on integrals of the motion. For a free mass the integrals of motion are the momentum and energy. For an oscillator they are the energy (and correspondingly the number of quanta), and also the quadrature amplitudes of equation (4.1). The quadrature amplitude of a traveling electromagnetic wave, in fact, was the first observable to be measured in a QND way.[28]

For an integral of the motion a sequence of repeated QND measurements can be useful, e.g., for detecting the influence of some external force that is acting on the object. In the absence of such an external force, the observable is conserved both during the measurement (because of the absence of a back-action perturbation on it) and between measurements (because the observable is an integral of the motion). Thus, perturbations of the object by previous QND measurements have no influence on the subsequent measurements. Correspondingly, the sensitivity to an external force is determined solely by the precision of the QND measurements, and by increasing the precision one in principle can increase the sensitivity without limit. For details see chapter VII.

A continuous QND monitoring of some observable is a natural limit of a sequence of many coherent QND measurements. For the position of a free mass or the generalized coordinate of an oscillator, any continuous monitoring will in principle be constrained by the standard quantum limit because the back action of the probe on the canonically conjugate momentum, followed by the evolutionary influence of the momentum on the coordinate, inevitably must affect the coordinate. For integrals of the motion, no such limits exist. The continuous monitoring of "ordinary" observables will be discussed in chapter VI, and continuous monitoring of integrals of the motion, in chapter VIII.

In concluding this section, we shall discuss one more example of a thought experiment for a QND measurement of a resonator's energy. The idea is simple: one can determine the energy in the resonator by weighing it. The weight of the electromagnetic energy E is

$$F_{e.m.} = \frac{E}{c^2} g \, ,$$

where g is the earth's gravitational acceleration. This force will change the momentum of a scale during the measurement time τ by the amount

$$\delta P = F_{\text{e.m.}}\tau = \frac{Eg\,\tau}{c^2} \ .$$

Thus, if the initial uncertainty of the scale's momentum was ΔP, then the electromagnetic energy can be measured in the best case with an accuracy

$$\Delta E_{\text{measure}} = \frac{c^2 \Delta P}{g\,\tau} \ .$$

As in our previous example, this measurement of the energy must be accompanied by a perturbation of the phase of the resonator's oscillations, a perturbation that, according to the uncertainty relation (4.15), must be at least as large as

$$\Delta\phi_{\text{perturb}} = \frac{\hbar\omega/2}{\Delta E_{\text{measure}}} = \frac{\hbar}{2\Delta P} \cdot \frac{g\,\omega\tau}{c^2} = \Delta y \frac{g\,\omega\tau}{c^2} \ .$$

Here $\Delta y = \hbar/(2\Delta P)$ is the uncertainty in the height of the resonator on the scale during the measurement. There will, indeed, be such a perturbation if during the measurement the frequency of the resonator has an uncertainty

$$\Delta\omega = \frac{\Delta\phi_{\text{perturb}}}{\tau} = \frac{g\,\omega}{c^2}\Delta y \ .$$

Thus, the frequency of the resonator must depend on its height in the gravitational field:

$$\omega = \omega_o(1 + gy/c^2) \ ,$$

where ω_o is the frequency at height $y = 0$.

By this analysis we have derived the famous formula from relativity theory, which describes the redshift of any frequency of oscillation in a uniform gravitational field—and we have done so not using directly the concepts of relativity theory, but instead using mainly the uncertainty principle.

4.3 The ponderomotive probe for energy

The ponderomotive probe, due to severe technical problems in its realization, is more a thought experiment than a foundation for a real measuring device. However, it can be regarded as a simple model for schemes of QND measurements for electromagnetic energy (see chapters XI and XII). Due to this, we shall analyze it in detail in this section.

Suppose that we have an electromagnetic resonator of the type shown in Fig. 4.1, one mode of which is excited with an energy E. Before the experiment its energy was known approximately, to within an uncertainty ΔE_{init}. On the resonator's right wall (as shown in Fig. 4.1), the electromagnetic pressure exerts the force $F = E/d$ [equation (4.10)]. The length d in this case is approximately the distance between the left and right walls. The measurement consists of a smooth motion of the right wall (during a time much longer than the electromagnetic period, so as to avoid parametrically exciting the resonator), followed by a measurement of the increase in the wall's momentum, $\delta P = E\tau/d$ [equation (4.11)].

Suppose that an "ordinary" method is used to measure the momentum, so its precision is constrained by the standard quantum limit $\sqrt{\hbar m/2\tau}$, where m is the mass of the wall [cf. equation (1.22) in section 1.4]. From this limit we obtain the following minimum value for the error of the measured energy:

$$\Delta E_1 = \frac{d}{\tau}\sqrt{\frac{\hbar m}{2\tau}} \ . \tag{4.16}$$

The larger is τ and the smaller are m and d, the smaller will be ΔE_1. However, one must take into account the fact that the resonator's electromagnetic field is a source of dynamical rigidity

$$K = E/d^2$$

for the mass's motion. This rigidity produces on the wall an unknown force

$$F' = K\,\Delta x \ ,$$

where

$$\Delta x = \sqrt{\hbar\tau/2m}$$

is the standard-quantum-limit uncertainty in the wall's position [cf. equation (1.23)]. In principle the rigidity can be compensated by some sort of externally applied spring, but only with an accuracy up to the unknown amount $\Delta E/d^2$. Thus, we may estimate the minimum size of the unknown force to be

$$F' = \frac{\Delta E_{init}}{d^2}\sqrt{\frac{\hbar\tau}{2m}} \ . \tag{4.17}$$

This unknown force produces an additional error in the measured value of the energy,

$$\Delta E_2 = F'd = \frac{\Delta E_{init}}{d}\sqrt{\frac{\hbar\tau}{2m}} \ . \tag{4.18}$$

This error ΔE_2, by contrast with ΔE_1, is larger the longer is τ and the smaller are m and d. The minimum total error occurs when $\Delta E_1 = \Delta E_2$. From equations (4.16) and (4.18) we see that to achieve this minimum we must choose the parameters such that

$$md^2 = \Delta E_{init} \cdot \tau^2 , \qquad (4.19)$$

and the error is then

$$\Delta E = [(\Delta E_1)^2 + (\Delta E_2)^2]^{1/2} = \sqrt{\frac{\Delta E_{init} \cdot \hbar}{\tau}} . \qquad (4.20)$$

In the worst case, when we know almost nothing about the energy before the measurement, we will have

$$\Delta E_{init} \simeq E , \qquad (4.21)$$

and correspondingly the minimum error in the measurement will be

$$\Delta E = \sqrt{\frac{E\hbar}{\tau}} = \frac{\Delta E_{SQL}}{\sqrt{\omega\tau}} . \qquad (4.22)$$

Here ΔE_{SQL} is the standard quantum limit for the energy [equation (1.25)], and ω is the frequency of the excited mode. Equation (4.22) shows that, if the measurement is made slowly so $\omega\tau \gg 1$, then the standard quantum limit can be beaten by a large factor, $\Delta E \ll \Delta E_{SQL}$, and one can even measure the energy with an accuracy better than one quantum, $\Delta E \ll \hbar\omega$.

Let us emphasize that equation (4.22) is not the fundamental limit of this ponderomotive method of energy measurement, since it was based on a very large initial uncertainty of the energy [equation (4.21)]. After a measurement with this large initial uncertainty, one knows the energy much more accurately, and one can use this new knowledge to more accurately compensate the electromagnetic rigidity and then repeat the measurement. The precision of the second measurement will be determined by equation (4.20) with ΔE_{init} equal to the error (4.22) of the first measurement. The result will be

$$\Delta E = E^{1/4}(\hbar/\tau)^{3/4} .$$

By repeating the measurement again and again, and each time compensating the electromagnetic rigidity with higher accuracy, we obtain in the limit

$$\Delta E = \frac{\hbar}{\tau} . \qquad (4.23)$$

In conclusion it is important to remark that the issue of the ultimate limit on the accuracy of an energy measurement is still open. The traditional viewpoint is an ultimate limit governed by Bohr's uncertainty relation

$$\Delta E \cdot \tau \geq \frac{\hbar}{2} \tag{4.24}$$

[which agrees in order of magnitude with (4.23).] However, this limit of accuracy has long been controversial, and recently schemes of measurement have been proposed that beat this limit.[29] On the other hand, there are doubts as to whether these schemes can be realized physically. In any event, when the energy of an oscillator is measured, then the following inequality must always be satisfied:

$$\Delta E \geq \frac{\hbar \omega}{2 \tau \Delta \omega} , \tag{4.25}$$

where $\Delta \omega$ is the uncertainty in the oscillator's frequency during the measurement. This inequality follows from the fact that the perturbation in the phase during the measurement,

$$\Delta \phi_{perturb} = \Delta \omega \tau ,$$

cannot be smaller than $\hbar \omega / 2 \Delta E$ [see equations (4.14) and (4.15)]. All methods of measuring electromagnetic energy that are regarded today as physically realistic obey the inequality $\Delta \omega \ll \omega$, and correspondingly they must produce errors

$$\Delta E \gg \frac{\hbar}{2 \tau} . \tag{4.26}$$

4.4* Criteria for QND measurements

To verify directly, by a theoretical calculation, that some specific measuring procedure will be a QND one is, as a rule, a sophisticated task. Thus, it is helpful to have some simple criteria that guarantee a procedure to be QND.

Relations (2.32) and (2.37) reveal that the operator $\hat{\Omega}(\tilde{q})$, which describes the back action of the probe during the measurement, must commute with the measured observable in order to avoid all back action on it. In other words,

$$[\hat{q}, \hat{\Omega}(\tilde{q})] = 0 \tag{4.27}$$

is a compact mathematical description of the definition of a QND measurement. Here $\tilde{q}$ is any result that can be obtained from the measurement.

By substituting expression (3.27) for $\hat{\Omega}(\tilde{q})$ into this formula, we obtain

$$<\tilde{q}\,|\,[\hat{q},\hat{U}]|\,\psi> = 0 , \tag{4.28}$$

where we recall that $|\tilde{q}>$ is the eigenstate of the probe corresponding to the measured value $\tilde{q}$ of the object's observable, $|\psi>$ is the probe's initial state, and $\hat{U}$ is the operator that generates the joint evolution of the probe and the object, from the beginning of the first step of the measurement to the end. Since the relation (4.28) must be satisfied identically for any value of $\tilde{q}$ that might be obtained in the measurement, it is equivalent to

$$[\hat{q},\,\hat{U}]|\,\psi> = 0 . \tag{4.29}$$

The physical meaning of this condition becomes more evident when one multiplies on the left by the operator $\hat{U}^{\dagger}$:

$$(\hat{U}^{\dagger}\hat{q}\,\hat{U} - \hat{q})|\,\psi> = 0 . \tag{4.30}$$

The product $\hat{U}^{\dagger}\hat{q}\,\hat{U}$ is exactly the value of the operator $\hat{q}$ in the Heisenberg picture at the end of the interaction between the probe and the object, and thus

$$\hat{U}^{\dagger}\hat{q}\,\hat{U} - \hat{q} \tag{4.31}$$

is the Heisenberg-picture change in the measured observable, produced by the interaction of the object and the probe.

Condition (4.30) for a QND measurement can be satisfied, in principle, by two methods. First, one can insist that the difference (4.31) vanish, i.e. that in the Heisenberg picture the measured observable must return to its initial value after the measurement. Second, if the observable does not so return, then one can choose the initial state of the probe to be an eigenstate of the difference (4.31).

The second method has not been developed at all, either theoretically or experimentally, despite the fact that it has some rather interesting possibilities. (For example, nothing in principle prevents the existence of a measurement scheme in which, by simply changing the initial state of the probe but leaving its coupling to the object unchanged, one can make successive QND measurements of different, *noncommuting* observables.)

The second method is usually ignored, and instead the first alone is regarded as the criterion for a QND measurement. This criterion says that the evolution operator must commute with the measured observable

$$[\hat{q},\,\hat{U}] = 0 . \tag{4.32}$$

The reason is the difficulty of preparing the quantum probe in a definite initial state. This is a complicated experimental task; it imposes the necessity to build some third quantum device to interact with the probe and thereby prepare its state before the measurement begins. In addition, this interaction and the initial state of this third quantum device must be chosen in a specific way, and so on. The condition (4.32) constrains more severely the interaction of the quantum probe with the object, but it is insensitive to the probe's initial state. Thus, in terms of practical realization, condition (4.32) is more attractive.

To verify whether condition (4.32) [or (4.30)] is satisfied requires a knowledge of the operator $\hat{U}$, and this in turn requires solving the evolution equations for the coupled quantum probe and measured object. This is generally a difficult task. For this reason, instead of the necessary and sufficient condition (4.32) for a measurement to be QND, a different, sufficient (but not necessary) condition is widely used: the measured quantity must be an integral of the motion for the coupled probe and object, i.e. it must not only return to its initial value (in the Heisenberg picture) at the end of the interaction, but it must remain constant throughout the interaction. From the equation of motion in the Heisenberg picture, it is easy to show that this sufficient condition is equivalent to the relation

$$i\hbar \frac{\partial \hat{q}}{\partial t} + [\hat{q}, \hat{H}] = 0 \; , \qquad (4.33)$$

where $\hat{H}$ is the Hamiltonian of the coupled system. If the measured observable has no explicit time dependence,

$$\frac{\partial \hat{q}}{\partial t} \equiv 0 \; ,$$

then condition (4.33) reduces simply to a demand that $\hat{q}$ commute with the Hamiltonian

$$[\hat{q}, \hat{H}] = 0 \; . \qquad (4.34)$$

It is evident that the condition (4.33) [or (4.34)] is more severe than condition (4.32). If it is satisfied, then the measurement is QND for any duration of the interaction between the meter and the object; in other words, the more general condition (4.32) is satisfied precisely, independently of the duration of the measurement. If one imposes only (4.32) for some chosen duration of measurement, then (4.33) may be violated during the measurement, and even (4.32) may fail if one switches to some other measurement duration.

The Hamiltonian $\hat{H}$ is typically of the form

$$\hat{H} = \hat{H}_{obj} + \hat{H}_{probe} + \hat{H}_I \; ,$$

where $\hat{H}_{\text{obj}}$ is the object's free Hamiltonian, $\hat{H}_{\text{probe}}$ is the Hamiltonian of the free quantum probe, and $\hat{H}_I$ is their interaction Hamiltonian. If the measured observable $\hat{q}$ is an integral of the object's free motion, then it satisfies

$$i\hbar\frac{\partial \hat{q}}{\partial t} + [\hat{q}, \hat{H}_{\text{obj}}] = 0 . \tag{4.35}$$

From this and from the obvious fact that

$$[\hat{q}, \hat{H}_{\text{probe}}] = 0 ,$$

it follows that when $\hat{q}$ is an integral of the free motion, the sufficient condition (4.33) for a measurement to be QND boils down to the condition that the measured observable must commute with the interaction Hamiltonian[*]

$$[\hat{q}, \hat{H}_I] = 0 . \tag{4.36}$$

---

[*]For another viewpoint on the formal theory of QND measurements, see Ref. 72.

V   Linear measurements

5.1 The measurement process and the uncertainty relation

In the unstarred sections of previous chapters, the back action of the measuring device on the measured object was analyzed by simply invoking the Heisenberg uncertainty relation for the perturbation and the measurement error. Is this simple method of analysis justified?

The measurement process does not show up in the standard formulation of the uncertainty relations. Instead, the standard formulation, which follows from the fundamental postulates of quantum mechanics (section 2.1), couples the intrinsic rms uncertainties of observables as defined by the quantum state. Nevertheless, the examples of indirect measurements discussed above (the Heisenberg microscope and Doppler speed meter in chapter I, and the electron probe in chapter III) exhibit a tight coupling between the measurement process and the uncertainty relations. In particular, both the measurement error and the perturbation of the object can be traced to the uncertainty properties of the *probe's* initial state, and from the fact that this initial state is subject to the uncertainty relations, one can show that the measurement error and perturbation are also subject to it.

Is it also possible, starting from the uncertainty relations as properties of the *object's* quantum state, to obtain the uncertainty relations for the measurement process? To answer this question, let us discuss the following thought experiment: We suppose that a generalized coordinate of some quantum object is measured, and that before the measurement the object was in a state with a very well defined momentum. The standard

Heisenberg relation dictates, then, a large initial uncertainty for the coordinate; i.e., before the measurement, very little is known about the coordinate. Then, after the measurement, the only information the experimenter will have about the coordinate is that obtained in the measurement, and thus the uncertainty of the coordinate in the object's final state Δx_{final} will be equal to the measurement error $\Delta x_{\text{measure}}$. At the same time, the uncertainty of the momentum, ΔP_{final}, in the object's final state will be equal to the measurement-induced perturbation, $\Delta P_{\text{perturb}}$ (because the initial momentum uncertainty was negligible). Therefore, from the Heisenberg relation for the object's final state

$$\Delta x_{\text{final}} \cdot \Delta P_{\text{final}} \geq \frac{\hbar}{2}$$

follows the uncertainty relation for the measurement

$$\Delta x_{\text{measure}} \cdot \Delta P_{\text{perturb}} \geq \frac{\hbar}{2} . \tag{5.1}$$

It is important to note that this argument relied crucially on the choice of the object's initial state as one with well-defined momentum. However, if the measurement process is so designed that it satisfies the key condition that *the error in the coordinate measurement and the perturbation of the momentum do not depend on the object's initial state*, then the uncertainty relation (5.1) will follow for any initial state. In this case, the entire process of measurement can be represented by two simple linear formulae:

$$\tilde{x} = x_{\text{init}} + \delta x_{\text{measure}} , \quad P_{\text{final}} = P_{\text{init}} + \delta P_{\text{perturb}} . \tag{5.2}$$

Here $\tilde{x}$ is the result of the coordinate measurement (the output signal of the probe), x_{init} and P_{init} are the initial values of the coordinate and momentum; and $\delta x_{\text{measure}}$ and $\delta P_{\text{perturb}}$ are random values that do not depend on the initial state of the object, and thus have no correlations with x_{init} and P_{init}, and thus have uncertainties (standard deviations) $\Delta x_{\text{measure}}$ and $\Delta P_{\text{perturb}}$ that satisfy the uncertainty relation (5.1) and depend only on the initial state of the probe.

Measurements of this type are called linear. Most of the examples discussed above belong to this class of measurements. However, it is not difficult to give examples that are nonlinear: Suppose that the energy E of a harmonic oscillator is measured, and we want to know the perturbation of its generalized coordinate x. The uncertainty relation for the energy and coordinate has the form

$$\Delta E \cdot \Delta x \geq \frac{\hbar |<P>|}{2m} , \tag{5.3}$$

where $<P>$ is the expectation value of the oscillator's momentum and m is its mass. Notice that the right-hand side of this inequality is not a constant, as it is, e.g., in (5.1), but instead is a quantity that depends on the oscillator's quantum state. If the oscillator is in a state with $<P> = 0$, then from this inequality one might think that there exists a scheme to measure the energy that does not perturb the coordinate. However, no such scheme exists; a more complete analysis shows that there is always a perturbation. Correspondingly, the class of linear measurements includes only those for which the uncertainty relations of all observables of interest (the measured ones and those whose perturbations one wishes to know) have only constants on the right-hand side.

As a second example, let us discuss the measurement of a particle's transverse (y) coordinate using the sieve of cells of section 2.4 and Fig. 2.1. The perturbation of the particle's y momentum in this example is of order $\hbar/\Delta y$. Suppose that the particle's initial state was such that the experimenter could predict in advance which cell the particle will traverse; i.e., the initial wave function was nonzero only within the boundary of that cell. Then the measurement will not change the particle's state. This can be proved rigorously using the mathematical formulation of the postulate of reduction as given in section 2.5, or by a direct solution for the diffraction of the particle's wave function. Thus, in this measurement scheme there exist initial states for the particle for which the measurement gives the observer no new information, and the particle is left unperturbed.

Thus, a simple analysis of the back action based on the uncertainty relation [equations (5.1) and (5.2)] is not always completely correct. In general such simple analyses are useful only for qualitative, semiclassical estimates, and to obtain fully correct results, one must use much more complicated methods based directly on the postulate of reduction [equations (2.23), (2.30), (3.25), and (3.27)].

It is important to emphasize that, from a practical point of view, the most important class of measurements is the class of linear ones. Linear measurements are closely connected to linear systems (those for which the equations of motion for the generalized coordinates and momenta are linear; for example, a free particle and a harmonic oscillator). More specifically: Real experiments generally involve not just one but several coherent measurements (for example of a generalized coordinate), or a continuous monitoring, which can be regarded as a sequence of a large number of separate measurements. Such experiments are linear measurements if the uncertainty relation for the observable at two different moments of time has a right-hand side that is independent of the measured object's state. One example is the position of a free particle, for which the two-time uncertainty relation reads

$$\Delta x(t)\cdot\Delta x(t') \geq \frac{\hbar}{2m}|t-t'| \ , \tag{5.4a}$$

where m is the particle's mass; another example is the generalized coordinate of an oscillator for which

$$\Delta x(t)\cdot\Delta x(t') \geq \frac{\hbar}{2m\omega}\sin(\omega|t-t'|) \ , \tag{5.4b}$$

where ω is the eigenfrequency. It is worth noting that all sensors for electromagnetic fields, except photon counters and certain schemes for QND measurements of energy, produce measurements that are linear in this sense, at least so long as the measured quantities are small enough (which, of course, is the regime of interest for quantum physics).

5.2* Measurement accuracy and perturbations for linear measurements

Let us discuss the measurement of some observable q. We want to know how some other observable P is perturbed if the measurement is linear. We shall base our analysis on expression (2.30) for the object's final state, and shall assume that the measurement is QND so the operator $\hat{\Omega}(\tilde{q})$ that appears in (2.30) commutes with the measured quantity:

$$[\hat{\Omega}(\tilde{q}),\hat{q}] = 0 \ . \tag{5.5}$$

This permits us to write $\hat{\Omega}$ in the form

$$\hat{\Omega}(\tilde{q}) = \int_{-\infty}^{+\infty} |q\!>\!\Omega(\tilde{q},q)\!<\!q|\,dq \ , \tag{5.6}$$

where $\Omega(\tilde{q},q)$ is some (in general complex) function that is normalized to unity

$$\int_{-\infty}^{+\infty} |\Omega(\tilde{q},q)|^2 d\tilde{q} = 1 \ . \tag{5.7}$$

The probability distribution for the results of the measurement, according to equations (2.26), (2.31), and (5.6), is the following:

$$w(\tilde{q}) = \int_{-\infty}^{+\infty} w_{\text{measure}}(\tilde{q}\,|\,q)\cdot w_{\text{init}}(q)\,dq \ , \tag{5.8}$$

where

$$w_{\text{init}}(q) = <\!q\,|\,\hat{\rho}_{\text{init}}\,|\,q\!> \tag{5.9}$$

is the probability distribution for the measured quantity in the initial state, and

$$w_{\text{measure}}(\tilde{q} \mid q) = |\Omega(\tilde{q}, q)|^2 \qquad (5.10)$$

is the conditional probability distribution, i.e. the probability that the result of the measurement will be $\tilde{q}$ when the true value of the measured observable is q. The form of this second distribution is determined by the properties of the measuring device, and its variance is, by definition, the measurement error.

The assumption of linearity implies that this measurement error is independent of q. It is reasonable to require also that the measurement be nonbiased, i.e., that the mean value of $\tilde{q}$ be equal to q:

$$\int\limits_{-\infty}^{+\infty} \tilde{q} w_{\text{measure}}(\tilde{q} \mid q) d\tilde{q} = q \ . \qquad (5.11)$$

From these two conditions it follows that the function w_{measure} depends only on the difference $\delta q = \tilde{q} - q$ of these two arguments, and in addition that

$$\int\limits_{-\infty}^{+\infty} \delta q \cdot w_{\text{measure}}(\delta q) d(\delta q) = 0 \ . \qquad (5.12)$$

Equation (5.8) in this case has the form of a convolution:

$$w(\tilde{q}) = \int\limits_{-\infty}^{+\infty} w_{\text{measure}}(\tilde{q} - q) w_{\text{init}}(q) dq \ . \qquad (5.13)$$

As is known from the classical theory of probability, this convolution describes the statistics of the sum of two independent random variables q and δq,

$$\tilde{q} = q + \delta q \ . \qquad (5.14)$$

The random variable q with the probability distribution $w_{\text{init}}(q)$ is the "true" value of the measured observable, and δq is an additional contribution produced by the measuring device. This additional contribution is governed by the probability distribution $w_{\text{measure}}(\delta q)$. According to (5.12), the mean value of δq is zero and its variance is given by

$$(\Delta q)^2 = \int\limits_{-\infty}^{+\infty} (\delta q)^2 w_{\text{measure}}(\delta q) d(\delta q) \ . \qquad (5.15)$$

The quantity Δq is the rms measurement error.

Next we shall calculate the perturbation of the observable P. According to equation (2.30), the probability distribution for P in the object's final state (presuming that the result of the measurement was $\tilde{q}$) is

$$W(P \mid \tilde{q}) = \frac{1}{w(\tilde{q})} <P \mid \hat{\Omega}(\tilde{q}) \hat{\rho}_{\text{init}} \Omega^{\dagger}(\tilde{q}) \mid P>. \tag{5.16}$$

This probability distribution depends both on the perturbation of the object during the measurement, and on the information obtained from the measurement. (In particular, the information about q influences the probability distribution for P if q and P were correlated in the object's initial state.)

To separate out the effects of the perturbation in a pure form, one must average this distribution over all possible results of the measurement:

$$W(P) = \int_{-\infty}^{+\infty} W(P \mid \tilde{q}) w(\tilde{q}) d\tilde{q} = D \int_{-\infty}^{+\infty} <P \mid \hat{\Omega}(\tilde{q}) \hat{\rho}_{\text{init}} \hat{\Omega}^{\dagger}(\tilde{q}) \mid P> d\tilde{q}$$

$$= \int_{-\infty}^{+\infty} <P \mid q> \Omega(\tilde{q},q) <q \mid \hat{\rho}_{\text{init}} \mid q'> \Omega^{*}(\tilde{q},q') <q' \mid P> d\tilde{q} dq dq'. \tag{5.17}$$

The linearity of the measurement produces two substantial simplifications of this expression. First, the commutator of $\hat{q}$ and $\hat{P}$ is a c-number

$$[\hat{q},\hat{P}] = ik , \tag{5.18}$$

where k is some constant (e.g. for a generalized coordinate and its corresponding momentum, $k = \hbar$), and correspondingly

$$<P \mid q> = \frac{1}{\sqrt{2\pi k}} e^{-iqP/k} . \tag{5.19}$$

Second, the function Ω depends only on the difference of its two arguments, and this permits Ω to be represented in the form

$$\Omega(\tilde{q},q) = \frac{1}{2\pi k} \int_{-\infty}^{+\infty} e^{i(q-\tilde{q})P/k} \Phi(P) dP , \tag{5.20}$$

where $\Phi(P)$ is the Fourier transform of the function Ω. By substituting equations (5.19) and (5.20) into (5.17) and performing a rather long, but in principle simple, calculation, one can obtain

$$W(P) = \int_{-\infty}^{+\infty} \mid \Phi(P-P') \mid^{2} W_{\text{init}}(P') dP' , \tag{5.21}$$

where

$$W_{init}(P) = <P \mid \hat{\rho}_{init} \mid P> \qquad (5.22)$$

is the initial probability distribution for the observable P.

Expression (5.21), like (5.13), is a convolution of two probability distributions, one of which is determined solely by the object's initial state, and the second, solely by the measuring device. Correspondingly, the final value of the object's momentum in the case of a linear measurement can be expressed in the form

$$P_{final} = P_{init} + \delta P \ , \qquad (5.23)$$

where the random variable δP is the measurement-induced perturbation of P. The statistics of δP are governed by the probability distribution $\mid \Phi(\delta P) \mid^2$, and thus the rms perturbation is the standard deviation ΔP of this distribution.

Thus, the probability distributions of both δq and δP are uniquely determined by the function $\Omega(\tilde{q}, q)$, via equations (5.10) and (5.20). Note that these equations have the same structure as the relation that connect the wave function in the q-representation to the probability distributions for the observables q and P. Thus, the uncertainties in δq and δP, i.e., the measurement error Δq and the perturbation ΔP, are equal to the uncertainties of the observables q and P in a state with the "wave function" Ω. This implies that these Δq and ΔP satisfy the uncertainty relation

$$\Delta q \cdot \Delta P \geq \frac{k}{2} \ . \qquad (5.24)$$

The minimum uncertainty, as is well known, is achieved when the wave function has a Gaussian form. Thus, equality is achieved in (5.24) when the function Ω has the form

$$\Omega(\tilde{q}, q) = \frac{1}{[2\pi(\Delta q)^2]^{1/4}} \exp\left[-\frac{(\tilde{q}-q)^2}{4(\Delta q)^2} \right] \ . \qquad (5.25)$$

5.3* Sequences of linear measurements

The main purpose of this section is to prepare some mathematical apparatus for the next chapter's analysis of continuous, linear measurements. However, the results obtained in this chapter can be used alone for analyses of complicated sequences of linear measurements.

Suppose that the observables q_1, q_2, ... , q_N are measured consecutively. The change of the state of the measured object in each measurement is described by equation (2.30). We shall assume that the measurements are all QND and linear.

The first measurement transfers the object from the initial state ρ_{init} to the state

$$\rho_1(\tilde{q}_1) = \frac{1}{w_1(\tilde{q})} \hat{\Omega}_1(\tilde{q}_1)\,\rho_{init}\,\hat{\Omega}_1^\dagger(q_1) \,, \tag{5.26}$$

where $\tilde{q}_1$ is the result of the first measurement, $\hat{\Omega}_1$ is the operator that describes the nature of the measurement, and

$$w_1(\tilde{q}_1) = <\hat{\Omega}^\dagger(\tilde{q}_1)\,\hat{\Omega}(\tilde{q}_1)>_{init} \tag{5.27}$$

is the probability distribution for $\tilde{q}_1$. To shorten future equations we introduce the notation

$$<\hat{Q}>_{init} = \mathrm{Tr}(\hat{Q}\,\rho_{init}) \,,$$

where $\hat{Q}$ is any operator.

For the second measurement, similarly,

$$\rho_2(\tilde{q}_1,\tilde{q}_2) = \frac{1}{w(\tilde{q}_2|\tilde{q}_1)} \hat{\Omega}_2(\tilde{q}_2)\,\rho_1(\tilde{q}_1)\,\hat{\Omega}_2^\dagger(\tilde{q}_2)$$

$$= \frac{1}{w(\tilde{q}_1,\tilde{q}_2)}\hat{\Omega}_2(\tilde{q}_2)\,\hat{\Omega}_1(\tilde{q}_1)\,\rho_{init}\,\hat{\Omega}_1^\dagger(\tilde{q}_1)\hat{\Omega}_2^\dagger(\tilde{q}_2) \,, \tag{5.28}$$

where

$$w(\tilde{q}_2|\tilde{q}_1) = \mathrm{Tr}\left[\hat{\Omega}_2^\dagger(\tilde{q}_2)\,\hat{\Omega}_2(\tilde{q}_2)\,\rho_1(\tilde{q}_1)\right]$$

is the probability to obtain the result $\tilde{q}_2$ in the second measurement, if the first measurement gave $\tilde{q}_1$, and

$$w(\tilde{q}_1,\tilde{q}_2)=w(\tilde{q}_2|\tilde{q}_1)\,w(\tilde{q}_1)=<\hat{\Omega}_1^\dagger(\tilde{q}_1)\,\hat{\Omega}_2^\dagger(\tilde{q}_2)\,\hat{\Omega}_2(\tilde{q}_2)\,\hat{\Omega}_1(\tilde{q}_1)>_{init} \tag{5.29}$$

is the joint probability distribution for the results of the two measurements. Repeating this same calculation for the third and subsequent measurements, we obtain, after the entire sequence of N measurements is finished, the following state for the measured object:

$$\rho_N(\tilde{q}_1,...,\tilde{q}_N) = \frac{1}{w(\tilde{q}_1,...,\tilde{q}_N)}\hat{Z}_N(\tilde{q}_1,...,\tilde{q}_N)\,\rho_{init}\,\hat{Z}_N^\dagger(\tilde{q}_1,...,\tilde{q}_N) \,, \tag{5.30}$$

where

$$\hat{Z}_j(\tilde{q}_1,...,\tilde{q}_j) = \hat{\Omega}_j(\tilde{q}_j)\,\cdots\,\hat{\Omega}_1(\tilde{q}_1) \tag{5.31}$$

for any j in the range from 1 to N,

$$w(\tilde{q}_1,...,\tilde{q}_N) = <\hat{\Pi}(\tilde{q}_1,...,\tilde{q}_N)> \tag{5.32}$$

is the joint probability distribution for the measured results, and the operator

$$\hat{\Pi}(\tilde{q}_1,...,\tilde{q}_N) = \hat{Z}_N^\dagger(\tilde{q}_1,...,\tilde{q}_N)\hat{Z}_N(\tilde{q}_1,...,\tilde{q}_N) \qquad (5.33)$$

creates the decomposition of unity that describes the properties of the measurement procedure.

Thus far we have assumed that the measurements are made one immediately after the other. However, it is not difficult to generalize these results to the case of measurements separated by intervals of time during with the object evolves freely. Denote by $\hat{\boldsymbol{U}}_j$ the object's evolution operator between measurements j and $j+1$. Then it is easy to show that equations (5.30), (5.32), and (5.33) are unchanged, and equation (5.31) is modified as follows:

$$\hat{Z}_j(\tilde{q}_1,...,\tilde{q}_j) = \hat{\Omega}_j(\tilde{q}_j)\hat{\boldsymbol{U}}_{j-1}\hat{\Omega}_{j-1}(\tilde{q}_{j-1}) \cdots \hat{\boldsymbol{U}}_1\hat{\Omega}_1 . \qquad (5.34)$$

Since we have assumed that the measurements are all QND and linear, and because of equation (5.6), $\hat{\Omega}_j(\tilde{q}_j)$ has the form

$$\hat{\Omega}_j(\tilde{q}_j) = \Omega_j(\tilde{q}_j - \hat{q}_j) \qquad (5.35)$$

for all $j = 1,...,N$. From this it follows that

$$\hat{\Omega}_2(q_2)\hat{\boldsymbol{U}}_1 = \hat{\boldsymbol{U}}_1\hat{\boldsymbol{U}}_1^\dagger\Omega_2(\tilde{q}_2)\hat{\boldsymbol{U}}_1 = \hat{\boldsymbol{U}}_1\hat{\Omega}_2{}'(\tilde{q}_2) ,$$

where

$$\hat{\Omega}_2{}'(\tilde{q}_2) = \Omega_2(\tilde{q}_2 - \hat{q}_2{}') ,$$

and

$$\hat{q}_2{}' = \hat{\boldsymbol{U}}_1^\dagger\hat{q}_2\hat{\boldsymbol{U}}_1$$

is the value of the operator $\hat{q}_2$ in the Heisenberg picture. Exactly similarly,

$$\hat{\Omega}_3(\tilde{q}_3)\hat{\boldsymbol{U}}_2\hat{\boldsymbol{U}}_1 = \hat{\boldsymbol{U}}_2\hat{\boldsymbol{U}}_1\hat{\Omega}_3{}'(\tilde{q}_3) ,$$

where

$$\hat{\Omega}_3{}'(\tilde{q}_3) = \Omega_3(\tilde{q}_3 - \hat{q}_3{}') ,$$

$$\hat{q}_3{}' = \hat{\boldsymbol{U}}_1^\dagger\hat{\boldsymbol{U}}_2^\dagger\hat{q}_3\hat{\boldsymbol{U}}_2\hat{\boldsymbol{U}}_1 ,$$

and so on. Thus, the operator (5.34) can be expressed in the form

$$\hat{Z}_N(\tilde{q}_1,...,\tilde{q}_N) = \hat{\boldsymbol{U}}_N \cdots \hat{\boldsymbol{U}}_1\hat{Z}_N{}'(\tilde{q}_1,...,\tilde{q}_N) ,$$

where

$$\hat{Z}_j{}'(\tilde{q}_1,...,\tilde{q}_j) = \hat{\Omega}_j{}'(\tilde{q}_j) \cdots \hat{\Omega}_1{}'(\tilde{q}_1) , \quad \text{for} \quad j = 1,...,N . \qquad (5.36)$$

Substituting this last expression into equation (5.33), we obtain

$$w(\tilde{q}_1,...,\tilde{q}_N) = <\hat{Z}'_N{}^\dagger(\tilde{q}_1,...,\tilde{q}_N)\hat{Z}_N{}'(\tilde{Q}_1,...,\tilde{q}_N)>_{\text{init}} . \qquad (5.37)$$

Thus, this procedure can be reduced to the one discussed above with only one difference: the values of the measured observables in the Heisenberg picture are fixed at the moment of their measurements.

Equation (5.32) or (5.37) in principle gives a full description of the statistics of the results of a sequence of measurements. However, these equations can be used directly in only a few special cases. They are not applicable to a continuous measurement; and for a sequence of discrete measurements, the information incorporated into them is usually very excessive. Typically one does not need to know the full probability distributions for the results of the measurements. Instead, one typically needs only the first two moments of the distributions, i.e. the expectation values, the variances, and the cross correlations. Let us calculate them.

The expectation value for the jth measurement is

$$<\tilde{q}_j> = \int\limits_{\{\tilde{q}\}} w(\tilde{q}_1,...,\tilde{q}_N)\tilde{q}_j \cdot d\tilde{q}_1 \cdots d\tilde{q}_N . \tag{5.38}$$

By inserting relations (5.32) and (5.33) into (5.38) and using conditions (5.7) and (5.11), we can bring this into the form

$$<\tilde{q}_j> = \int\limits_{\{\tilde{q}\}} <\hat{Z}_{j-1}^{\dagger}(\tilde{q}_1,...,\tilde{q}_{j-1})\hat{q}_j\hat{Z}_{j-1}(\tilde{q}_1,...,\tilde{q}_{j-1})>_{\text{init}}d\tilde{q}_1 \cdots d\tilde{q}_{j-1} , \tag{5.39}$$

where by $\int\limits_{\{\tilde{q}\}}$ is meant an integral over all possible results of the measurements.

To bring this into a more physically clear form, we perform further manipulations which begin with the obvious relation

$$\hat{q}_j\hat{\Omega}_{j-1} = \hat{\Omega}_{j-1}\hat{q}_j + [\hat{q}_j,\hat{\Omega}_{j-1}] . \tag{5.40}$$

Linearity of the measurements implies that the commutators of all the measured observables are c-numbers:

$$[\hat{q}_j,\hat{q}_l] = ik_{jl} , \tag{5.41}$$

where the constants k_{jl} form an $N \times N$ matrix. From equations (5.35) and (5.41) it follows that

$$[\hat{q}_j,\hat{\Omega}_l] = \frac{\partial\Omega_l(\tilde{q}_l-\hat{q}_l)}{\partial\hat{q}_l} \cdot ik_{jl} . \tag{5.42}$$

By inserting expressions (5.42) and (5.40) into (5.39) and performing straightforward manipulations, one obtains

$$<\hat{q}_j> = \int\limits_{\{\tilde{q}\}} <\hat{Z}_{j-2}^{\dagger}(\tilde{q}_1,...,\tilde{q}_{j-2})\hat{q}_j\hat{Z}_{j-2}(\tilde{q}_1,...,\tilde{q}_{j-2})>_{\text{init}}d\tilde{q}_1 \cdots d\tilde{q}_{j-2}$$

$$+ ik_{j-1\,j}\int\limits_{-\infty}^{+\infty} \Omega_{j-1}^{*}(q)\frac{d\Omega_{j-1}(q)}{dq}dq . \tag{5.43}$$

Repeating the sequence of steps (5.40)—(5.43) $j-2$ times, we obtain, finally,

$$<\hat{q}_j> \ = \ <\hat{q}_j>_{\text{init}} + i\sum_{l=1}^{j-1}k_{lj}\int_{-\infty}^{+\infty}\Omega_l^*(q)\frac{d\Omega_l(q)}{dq}dq \ . \tag{5.44}$$

The physical meaning of this relation should be evident: The mean value for the result of the jth measurement is equal to the initial expectation value of the jth observable, plus the sum of the perturbations of its mean in the first $j-1$ measurements. Notice that, if, for some l, Ω_l is a real function, then the lth measurement does not perturb the mean values of the observables measured subsequently. Thus, it is evident that it is not necessary in principle for the mean values to be perturbed. This motivates us, so as to prevent subsequent calculations from becoming overly long, to assume henceforth that the mean values are left unperturbed by all the measurements, i.e. that all the functions Ω_j are real, and correspondingly all the operators $\hat{\Omega}_j$ are Hermitian.

Turn, next, to the second moment (mean square) of the result for the jth measurement. It is given by

$$<\tilde{q}_j^2> \ = \ \int_{\{\tilde{q}\}} w(\tilde{q}_1,...,\tilde{q}_N)\tilde{q}_j^2 d\tilde{q}_1\cdots d\tilde{q}_N$$

$$= \ \int_{\{\tilde{q}\}}<\hat{Z}_{j-1}^\dagger(\tilde{q}_1,...,\tilde{q}_{j-1})[\hat{q}_j^2+(\Delta q_j)^2]\hat{Z}_{j-1}(\tilde{q}_1,...,\tilde{q}_{j-1})>_{\text{init}}d\tilde{q}_1\cdots d\tilde{q}_{j-1} \ ,$$

$$\tag{5.45}$$

where

$$\Delta q_j \ = \ \left[\int_{-\infty}^{+\infty}q^2|\Omega_j(q)|^2 dq\right]^{1/2} \tag{5.46}$$

is the mean square error of the jth measurement. From equations (5.40) and (5.42) it follows (assuming that the operators $\hat{\Omega}_j$ are all Hermitian) that

$$\hat{\Omega}_{j-1}\hat{q}_j^2\hat{\Omega}_{j-1} \ = \ \hat{q}_j\hat{\Omega}_{j-1}^2\hat{q}_j + \left[\frac{\partial\Omega_{j-1}(\tilde{q}-\hat{q}_{j-1})}{\partial\hat{q}_{j-1}}\right]^2 k_{j-1\,j} \ .$$

By substituting this equation into (5.46), we obtain

$$<\tilde{q}_j^2> \ = \ \int_q<\hat{Z}_{j-2}^\dagger(\tilde{q}_1,...,\tilde{q}_{j-2})\hat{q}_j^2\hat{Z}_{j-2}(\tilde{q}_1,...,\tilde{q}_{j-2})>_{\text{init}}d\tilde{q}_1\cdots d\tilde{q}_{j-2}$$

$$+ (\Delta q_j)^2 + k_{j-1\,j}^2\int_{-\infty}^{+\infty}\left[\frac{d\Omega_{j-1}(q)}{dq}\right]^2 dq \ . \tag{5.47}$$

By repeating these manipulations $j-2$ times, we obtain the final form for the mean square result of the jth measurement:

$$<\tilde{q}_j^2> = <\hat{q}_j^2>_{\text{init}} + (\Delta q_j)^2 + \sum_{l=1}^{j-1} k_{lj}^2 \int_{-\infty}^{+\infty} \left[\frac{d\Omega_l(q)}{dq} \right]^2 dq \ . \quad (5.48)$$

The physical interpretation of this result is rather simple. The mean square of the results of each measurement in the sequence is a sum of three contributions: the initial mean square of the measured quantity [the first term in (5.48)], the measurement error (the second term), and a perturbation produced by the preceding $j-1$ measurements.

Our final task is to compute the cross correlation of the results of two different measurements. We omit the details of the calculation, since they are absolutely similar to those above, and only write down the result. For $j \neq l$,

$$<\tilde{q}_j \tilde{q}_l> = <\hat{q}_j \circ \hat{q}_l>_{\text{init}} + \sum_{n-1}^{\min(j,l)} k_{jn} k_{ln} \int_{-\infty}^{+\infty} \left[\frac{d\Omega_n(q)}{dq} \right]^2 dq \ . \quad (5.49)$$

Here $\hat{q}_j \circ \hat{q}_l$ is the symmetrized product, $\frac{1}{2}(\hat{q}_j \hat{q}_l + \hat{q}_l \hat{q}_j)$.

From equations (5.44), (5.48), and (5.49) follows an expression for a covariant matrix that describes the sequence of measurements. This matrix describes the likely spread of the results of a number of different experiments, and also the statistical correlations. More specifically, this matrix is a discrete analog of the correlation function for random processes:

$$B_{jl} \equiv <(\tilde{q}_j - <\tilde{q}_j>)(\tilde{q}_l - <\tilde{q}_l>)>$$

$$= B_{jl}^{\text{init}} + (\Delta q_j)^2 \delta_{jl} + \sum_{n=1}^{\min(j,l)} k_{jn} k_{ln} \int_{-\infty}^{+\infty} \left[\frac{\partial \Omega_n(q)}{\partial q} \right]^2 dq \ , \quad (5.50)$$

where the matrix

$$B_{jl}^{\text{init}} = <(\hat{q}_j - <\hat{q}_j>_{\text{init}}) \circ (\hat{q}_l - <\hat{q}_l>_{\text{init}})>_{\text{init}} \quad (5.51)$$

is determined by the object's initial state.

As was mentioned at the end of the last section, for a fixed measurement error, the perturbation is minimized if Ω has a Gaussian form. If this is the case for all measurements in the sequence, then (5.50) takes on the following form:

$$B_{jl} = B_{jl}^{\text{init}} + (\Delta q_j)^2 \delta_{jl} + \sum_{n=1}^{\min(j,l)} \frac{k_{jn} k_{ln}}{4(\Delta q_n)^2} \ .$$

VI Continuous linear measurements

6.1 Discrete and continuous measurements

All the examples of measurements so far discussed have in common the fact that the experimental output is either a single number or a finite set of numbers. However, in real measurements the output is often a record of some continuous function of time, from which one can get an understanding of the behavior of the measured quantity during some time interval. This type of measurement is called continuous. Correspondingly, measurements that give a discrete set of numbers can be called discrete.

Let us note that measurements are sometimes called continuous if they are made not instantaneously, but over a finite duration of time. Evidently, this is not a reasonable viewpoint; all real measurements last for a finite time. The dividing line between continuous and discrete measurements is in fact the character of the output signal: does it give information about the measured quantity only at some chosen moment of time, or does it permit a continuous monitoring of the quantity's time evolution.

The necessity to develop the quantum theory of continuous measurements arose in the 1960s, when devices for making continuous measurements were developed with sensitivities close to the quantum level (masers, parametric amplifiers, optical heterodyne receivers, ...). The "traditional" quantum theory of measurement, which was completed by the beginning of the 1970s, did not treat continuous measurements.

During the 1980s several different approaches to the analysis of continuous quantum measurements were developed.[30–32] Despite the substantial difference in the mathematical apparatus used in these approaches, most of them are close to each other in essence. Nearly all are based on the fact that there is no impassable wall between continuous measurements and sequences of discrete measurements. In fact, it is well known from the theory of information transfer that any continuous signal with a limited spectrum (and the spectrum of real signals is always limited) can be uniquely reconstructed from a discrete set of sampled values. It is necessary only that the time interval between samples be several times smaller than the shortest period present in the spectrum. Thus, a continuous measurement can be regarded as the limiting case of a sequence of discrete measurements, when the interval between the measurements is much smaller than the characteristic timescale on which the measured quantity changes.

The most difficult thing to compute correctly in this case is the perturbation of the measured object. Any attempt to compute it straightforwardly produces complicated multidimensional integrals (in the rigorous limit, infinite dimensional ones). None of the modern approaches permit one to obtain, in the general, nonlinear case, general relations that describe the properties of continuous measurements and that are practical to use.

However, a reasonably simple and general theory can be developed if one limits oneself to only linear measurements. This theory is the subject of this chapter. Several aspects of the theory of nonlinear continuous measurements will be discussed in the next chapter.

6.2 Uncertainty relations for continuous linear measurements

Let us discuss a simple thought experiment consisting of a sequence of measurements of the position $x(t)$ of some object. The measurements follow one after another with a constant separation time θ. We shall assume that the measurements are linear and thus can be described by equations (5.2). The error of each measurement $\Delta x_{measure}$ and the accompanying perturbation $\Delta P_{perturb}$ of the object's momentum are connected by the Heisenberg uncertainty relations (5.1).

The measuring device might be, for example, the Heisenberg microscope of section 1.3, in which the object is illuminated by a flux of photons. In this case

$$\Delta x_{measure} = \xi\lambda \, , \qquad \Delta P_{perturb} = \frac{\hbar}{2\xi\lambda} \, ,$$

where λ is the wavelength of the light and ξ is a factor of order unity that depends on the geometry of the microscope.

If the time interval θ is short enough, then the object's position x will not change substantially between measurements, and it will be reasonable to average the results of several successive measurements and thereby improve the measurement precision. If n measurements are averaged, then the position will be known with the precision

$$\Delta_\tau x = \frac{\Delta x_{measure}}{\sqrt{n}} = \Delta x_{measure}\sqrt{\frac{\theta}{\tau}}, \qquad (6.1)$$

where $\tau = n\theta$ is the averaging time.

Let us now reduce the interval θ between measurements to zero and increase the error of each measurement to infinity, in such a way that the quantity

$$S_x = (\Delta x_{measure})^2\cdot\theta$$

remains constant. In the case of the Heisenberg microscope, for example,

$$S_x = \xi^2\lambda^2\theta ,$$

so it is necessary to increase the wavelength while decreasing θ. The error of a single measurement will go up, but the precision during a fixed averaging time τ will remain unchanged:

$$\Delta_\tau x = \sqrt{\frac{S_x}{\tau}} . \qquad (6.2)$$

To clarify the physical meaning of the quantity S_x, let us discuss the first of the formulas (5.2). In the limit $\theta \to 0$, this formula takes the form

$$\tilde{x}(t) = x(t) + x_{fluct}(t) , \qquad (6.3)$$

where $\tilde{x}(t)$ is the output signal of the measuring device. Thus, the output signal is the sum of the input signal $x(t)$ and the noise $x_{fluct}(t)$ added by the meter. For this simple model the values of $x_{fluct}(t)$ at different moments of time are independent (because they originated from errors of *different* measurements in the pre-limit sequence). In other words, the noise is white. The quantity S_x is the spectral density of this noise.

We turn next to an analysis of the back action of the measuring device on the object. During the measurement, the object periodically (at time intervals θ) receives a random kick. The rms change of momentum due to the cumulative effect of these kicks is $\Delta P_{perturb}$ [cf. the second of the formulas (5.2)]. In other words, the momentum undergoes a "Brownian motion" (superimposed on regular changes due to the object's eigenevolution). The variance of the momentum increases in the manner of a diffusive process (as in Brownian motion): after a time τ the rms perturbation has grown to

$$\Delta_\tau P = \Delta P_{\text{perturb}} \cdot \sqrt{n} = \Delta P_{\text{perturb}} \cdot \sqrt{\frac{\tau}{\theta}} .$$

This equation can also be expressed as

$$\Delta_\tau P = \sqrt{S_F \tau} , \tag{6.4}$$

where

$$S_F = \frac{1}{\theta}(\Delta P_{\text{perturb}})^2 \tag{6.5}$$

is the diffusion coefficient. For the Heisenberg microscope,

$$S_F = \frac{\hbar^2}{4\xi^2\lambda^2\theta} . \tag{6.6}$$

If $\theta \rightarrow 0$ and $\lambda \rightarrow \infty$ in the same manner as above, then S_F remains constant. The collisions of the object with the photons become more frequent and the change of momentum in each collision becomes smaller in such a way that the mean square deviation of the momentum after a fixed time remains constant. In the limit, the separated kicks become a continuous, random force. The quantity S_F is the spectral density of that force.

By taking the product of equations (6.2) and (6.4) and noting that the quantities $\Delta x_{\text{measure}}$ and $\Delta P_{\text{perturb}}$ are connected by the uncertainty relation (5.1), we obtain the following simple inequality:

$$S_x S_F \geq \frac{\hbar^2}{4} . \tag{6.7}$$

The similarity between this inequality and the Heisenberg uncertainty relation is evident. This similarity is natural, because the inequality (6.7) is a direct consequence of the uncertainty relation for each of the measurements in the pre-limit sequence. For continuous linear measurements, equation (6.7) plays the same role as does the Heisenberg relation for discrete measurements: it establishes a universal, mutual connection between the accuracy of the monitoring and the perturbation of the monitored object.

Let us discuss one additional example of a continuous measurement. Suppose that one wishes to measure the angular coordinate ϕ of some object that rotates around its z-axis, e.g., a torsion pendulum. One can do so by attaching to the object a mirror, as shown in Fig. 6.1, and registering the angle of reflection of a beam of light that falls on the mirror (the same technique as is used in a galvanometer).

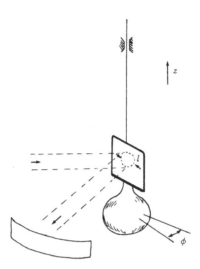

Fig. 6.1 Measurement of the angular coordinate of a torsion pendulum.

The precision of measurement in such an experiment is limited by diffraction of the reflected light beam. The resulting uncertainty in the measured angle, when just one photon is used, is

$$\Delta\phi_{diffr} = \frac{c}{\omega l} ,$$

where l is the size of the light spot on the mirror and ω is the frequency of the light. When the mirror rotates through an angle $\delta\phi$, the reflected beam rotates by $2\delta\phi$, so using one photon one makes a measurement error

$$\Delta\phi_{measure} = \frac{\Delta\phi_{diffr}}{2} = \frac{c}{2\omega l} .$$

Suppose that during the measurement time τ, n photons are reflected. (We presume that τ is sufficiently small that during it the mirror's angle does not change significantly.) Then the measurement error will be

$$\Delta_\tau\phi = \frac{\Delta\phi_{measure}}{\sqrt{n}} = \frac{c}{2\omega l}\sqrt{\frac{\theta}{\tau}} ,$$

where $\theta = \tau/n$ is the mean interval between photons.

To achieve a correct limit to a continuous monitoring of the angle ϕ, it is necessary to let $\theta \rightarrow 0$ and $\omega \rightarrow 0$, while holding the value of θ/ω^2 constant (i.e., it is necessary to use many low frequency photons). The accuracy of the monitoring then is described by the following formula, which is similar to (6.2):

$$\Delta_\tau \phi = \sqrt{\frac{S_\phi}{\tau}} \, ,$$

where the quantity

$$S_\phi = \frac{c^2 \theta}{4 \omega^2 l^2}$$

is the spectral density of the noise that is added to the measured quantity by the measuring device [see also equations (6.2) and (6.4)].

In this example, the back action of the measurement on the object is produced by light-pressure fluctuations on the mirror. This is a purely quantum effect. If the light beam were a classical wave, then it would be possible to produce identical light spots on the two sides of the mirror, with identically the same pressures, and the net pressure would vanish. However, each photon kicks the mirror at a random location, and thus its angular-momentum transfer is also random, with an uncertainty

$$\Delta L_{\text{perturb}} = \frac{2\hbar\omega}{c} \frac{l}{2} = \frac{\hbar\omega l}{c}$$

where $2\hbar\omega/c$ is the linear momentum that the photon gives to the mirror.

After n photons have been reflected from the mirror, the uncertainty of its angular momentum will have increased to

$$\Delta_\tau L = \frac{\hbar\omega l}{c} \sqrt{n} \ .$$

When we take the limit $\theta \to 0$ and $\omega \to 0$ in the same way as above, we can rewrite this equation in a form analogous to (6.4):

$$\Delta_\tau L = \sqrt{S_L \tau} \, ,$$

where

$$S_L = \frac{\hbar^2 \omega^2 l^2}{\theta c^2}$$

is the spectral density of the fluctuations of angular momentum produced by the light pressure. It is easy to show that the quantities S_ϕ and S_L satisfy the uncertainty relation

$$S_\phi \cdot S_L \geq \frac{\hbar^2}{4} \ . \tag{6.8}$$

In conclusion, we note that the examples we have discussed are the simplest of all possible examples. In other cases the inequality (6.7) requires generalization. In particular, there are situations in which

1. the accuracy of the monitoring changes with time;

2. the error of the monitoring and the back action are mutually correlated;

3. the measurement errors and/or the perturbations of the object at different moments of time are not independent.

In the remaining sections of this chapter these generalizations will be discussed.

6.3* Uncertainty relations for continuous linear measurements— rigorous analysis

The results of the last section can be generalized to the continuous, linear monitoring of any observable. We shall do so, starting from equation (5.50) of the last chapter, which describes the statistical properties of the results of a sequence of discrete, linear measurements. We suppose that the observable $q(t)$ is measured sequentially at the moments $t_j = j\theta$ ($j = 0, 1, 2, \dots$). Linearity guarantees that this observable's self-commutator (in the Heisenberg picture) is a c-number:

$$[\hat{q}(t), \hat{q}(t')] = iK(t, t') .$$

In this case, equation (5.50) can be written in the form

$$B_{jl} = B_{jl}^{\text{init}} + (\Delta q_j)^2 \delta_{jl} + \frac{1}{\hbar^2} \sum_{n=1}^{\min(j,l)} K(t_j, t_n) K(t_l, t_n) \sigma_n^2 , \qquad (6.9)$$

where

$$\sigma_n^2 = \hbar^2 \int_{-\infty}^{+\infty} \left[\frac{d\Omega_n(q)}{dq} \right]^2 dq \geq \frac{\hbar^2}{4(\Delta q_n)^2} . \qquad (6.10)$$

(Equality occurs when Ω is a Gaussian function.)

To make a correct transition to the limit of continuous monitoring it is convenient to use the following method. Consider the random variable

$$Q = \theta \sum_j \alpha(t_j) \tilde{q}(t_j) ,$$

where $\alpha(t)$ is an arbitrary, sufficiently smooth function (a function that changes little during a time θ.) The summation is taken over all the measurements in the sequence. The variance of Q is given by

$$(\Delta Q)^2 = \theta^2 \sum_{jl} \alpha(t_j) \alpha(t_l) B_{jl}$$

$$= \theta^2 \left[\sum_{jl} \alpha(t_j) \alpha(t_l) B_{jl}^{\text{init}} + \sum_j (\Delta q_j)^2 \alpha^2(t_j) \right]$$

$$+ \sum_{jl} \sum_{n < \min(j,l)} \alpha(t_j)\,\alpha(t_l){\cdot}K(t_j,t_n)\,K(t_l,t_n)\sigma_n^2 \Bigg] \; . \tag{6.11}$$

Since the summands do not change significantly when the indices change by unity, the sums in (6.11) can be replaced by integrals using the relation $\theta\sum \to \int$. This brings equation (6.11) into the following form:

$$(\Delta Q)^2 = \int\limits_0^{\tau_{\text{measure}}} \alpha(t)\alpha(t')B(t,t')dtdt'$$

$$= \int\limits_0^{\tau_{\text{measure}}} \alpha(t){\cdot}\alpha(t')B^{\text{init}}(t,t')dtdt' + \int\limits_0^{\tau_{\text{measure}}} \alpha^2(t)S_q(t)dt \tag{6.12}$$

$$+ \int\limits_0^{\tau_{\text{measure}}} dtdt' \int\limits_0^{\min(t,t')} dt''\alpha(t)\alpha(t')K(t,t'')K(t,t'')S_F(t'') \; ,$$

where τ_{measure} is the measuring time, $B(t,t')$ is the correlation function for the measuring device's output signal,

$$B^{\text{init}}(t,t') = \big<(\hat{q}(t){-}{<}(\hat{q}(t){>}_{\text{init}}){\circ}(\hat{q}(t'){-}{<}\hat{q}(t'){>}_{\text{init}})\big>_{\text{init}}$$

is the "unperturbed correlation function" that describes the statistical properties of the measured quantity in the absence of any influence from the measuring device, and

$$S_q(t_j) = \lim_{\theta \to 0} \theta{\cdot}(\Delta q_j)^2 \; , \quad S_F(t_j) = \lim_{\theta \to 0} \frac{\sigma_j^2}{\theta} \; . \tag{6.13}$$

Since $\alpha(t)$ is an arbitrary function, it follows from equation (6.13) that

$$B(t,t') = B^{\text{init}}(t,t') + S_q(t){\cdot}\delta(t{-}t') +$$

$$+ \frac{1}{\hbar^2} \int\limits_0^{\min(t,t')} K(t,t'')K(t',t'')S_F(t'')dt'' \; . \tag{6.14}$$

Now, the self-commutator of any generalized coordinate is related to the corresponding generalized susceptibility $\chi(t,t')$ by

$$\chi(t,t') = \begin{cases} 0 \; , & t < t' \\[2mm] -\dfrac{1}{\hbar}K(t,t') \; , & t > t' \; . \end{cases} \tag{6.15}$$

By inserting this relation into equation (6.14), we obtain

$$B(t,t') = B^{\text{init}}(t,t') + S_q(t){\cdot}\delta(t{-}t')$$

$$+ \int\limits_0^{\min(t,t')} \chi(t,t'')\chi(t',t'')S_F(t'')dt'' \; . \tag{6.16}$$

This relation helps to clarify the physical meaning of $S_F(t)$. It characterizes the strength of the random force that the measuring device exerts on the measured object. The correlation function of this back-action force is

$$B_F(t,t') = S_F(t) \cdot \delta(t-t') .$$

At the same time, $B_q(t,t')$ is the correlation function of the noise that is added to the output signal by the measuring device:

$$B_q(t,t') = S_q(t)\delta(t-t') .$$

Equation (6.10) implies the following uncertainty relation between $S_q(t)$ and $S_F(t)$:

$$S_q(t) \cdot S_F(t) \geq \frac{\hbar^2}{4} . \tag{6.17}$$

This generalizes the uncertainty relation (6.7) to the case of a monitoring of any observable, and the above analysis shows that this relation remains valid when the accuracy of the monitoring varies with time.

6.4* Linear, quantum 2N-pole systems

It is possible, in principle, to construct a theory of any continuous, linear measurement, using the methods of the last section. However, for the general case the calculations become inordinately cumbersome. Accordingly, in the remainder of this chapter we shall use a different approach, one that is physically less transparent but mathematically more simple.

Any system for continuously monitoring some quantity can be regarded as a black box that makes contact with the external world through two ports—or, in the language of electronics, a "four-pole system." One port (or pair of contact poles) is the input into the system; the other is the output, from which we read the results of the monitoring. Each port has its own generalized coordinate Z_j (with $j = 1, 2$), and the external action on it can be regarded as a generalized force $G_j(t)$. For a linear measuring system, the time evolution of the operator $\hat{Z}_j(t)$ in the Heisenberg picture is described by a linear equation of the form

$$\hat{Z}_j(t) = \hat{Z}_j^o(t) + \sum_{k=1}^{N} \int_{-\infty}^{+\infty} \chi_{jk}(t,t')G_k(t')dt' , \tag{6.18}$$

where N is the number of ports (in this particular case, $N = 2$), $\chi_{jk}(t,t')$ are the generalized susceptibilities of the system (an $N \times N$ matrix), and $\hat{Z}_j^o(t)$ are the values of the generalized coordinates in the absence of any external action. Since the system is linear, we can set $\langle \hat{Z}_j^0(t) \rangle \equiv 0$ without loss of generality, by redefining

$$\hat{Z}_{j\ \mathrm{new}}(t) = \hat{Z}_{j\ \mathrm{old}}(t) - <\hat{Z}^{o}_{j\ \mathrm{old}}(t)> \ .$$

The correlation matrix

$$B_{jk}(t,t') = <Z^{o}_{j}(t) \circ \hat{Z}^{o}_{k}(t')>$$

provides a quantitative measure of the system's internal fluctuations. There is an obvious connection between this correlation matrix and the matrix of susceptibilities. For a system in equilibrium, this connection is described by the fluctuation-dissipation theorem. However, the fluctuation-dissipation theorem breaks down for systems such as measuring devices that are far from equilibrium. We shall now search for an appropriate generalization of the fluctuation-dissipation theorem: a general relation connecting the correlation matrix and the matrix of susceptibilities of any linear system, including one not in equilibrium. In doing so, we shall allow the system's number N of inputs and outputs to be arbitrary, i.e. we shall analyze an arbitrary, linear, $2N$-pole system.

The Hamiltonian of such a system has the form

$$\hat{H} = \hat{H}_{o} - \sum_{j=1}^{N} G_{j}(t)\hat{Z}_{j} \ , \qquad (6.19)$$

where $\hat{H}_{o}$ is the unperturbed Hamiltonian. Using perturbation theory, one can derive the solution to this Hamiltonian's equations of motion for the operator $\hat{Z}_{j}$ (in the Heisenberg picture):

$$\hat{Z}_{j}(t) = \hat{Z}^{o}_{j}(t) + \frac{i}{\hbar} \sum_{k=1}^{N} \int_{-\infty}^{t} [\hat{Z}^{o}_{j}(t), \hat{Z}^{o}_{k}(t')] G_{k}(t')dt' + \dots$$

$$+ \left[\frac{i}{\hbar}\right]^{n} \sum_{k_{1},\dots,k_{n}=1}^{N} \int_{-\infty}^{t} dt_{1} \cdots \int_{-\infty}^{t_{n-1}} dt_{n} [\ \cdots\ [\hat{Z}^{o}_{j}(t), \hat{Z}^{o}_{k_{1}}(t_{1})],\dots,\hat{Z}^{o}_{k_{n}}(t_{n})]\cdot$$

$$\cdot G_{k_{1}}(t_{1}) \cdots G_{k_{n}}(t_{n}) + \dots \ . \qquad (6.20)$$

For a linear system, as one can see by comparing equations (6.17) and (6.20), this series stops after the first two terms. In order for this to happen, the commutators and self-commutators of the operators $\hat{Z}^{o}_{j}(t)$ must be c-numbers and not operators. As we have seen [cf. equation (6.15)], the commutators in this case are described by the $2N$-pole system's matrix of generalized susceptibilities:

$$\hat{\chi}_{jk}(t,t') = \begin{cases} 0 \ , & t \geq t' \\ \dfrac{i}{\hbar}[\hat{Z}^{o}_{j}(t), \hat{Z}^{o}_{k}(t')] \ , & t < t' \ . \end{cases} \qquad (6.21)$$

Consider, now, any operator $\hat{Q}$ with the form

$$\hat{Q} = \sum_{j=1}^{N} Q_j(t)\hat{Z}_j^o(t) , \qquad (6.22)$$

where $Q_j(t)$ are arbitrary complex functions. The expectation value of the product of this operator and its Hermitian conjugate evidently cannot be negative

$$<\hat{Q}^\dagger\hat{Q}> = \sum_{j,k=1}^{N} \int_{-\infty}^{+\infty} Q_j^*(t)Q_k(t)<\hat{Z}_j^o(t)\hat{Z}_k^o(t')> dt dt' \geq 0 . \qquad (6.23)$$

This expression can be brought into a more useful form with the aid of the identity

$$<\hat{Z}_j^o(t)\hat{Z}_k^o(t')> \equiv <\hat{Z}_j^o(t)o\hat{Z}_k^o(t')> + \frac{1}{2}<[\hat{Z}_j^o(t),\hat{Z}_k^o(t')]> \qquad (6.24)$$

$$= B_{jk}(t,t') - \frac{i\hbar}{2}[\chi_{jk}(t,t')-\chi_{kj}(t',t)] ,$$

where

$$\chi_{jk}(t,t') = <\hat{\chi}_{jk}(t,t')>$$

is the expectation value of the susceptibility. By inserting (6.24) into (6.23), we conclude that any linear quantum system must satisfy the following inequality:

$$\sum_{j,k} \int_{-\infty}^{+\infty} Q_j^*(t) Q_k(t') B_{jk}(t,t') dt dt'$$

$$\geq \frac{i\hbar}{2} \int_{-\infty}^{+\infty} Q_j^*(t) Q_k(t') [\chi_{jk}(t,t')-\chi_{kj}(t',t)] dt dt' . \qquad (6.25)$$

Because the functions $Q_j(t)$ are arbitrary, the inequality (6.25) guarantees the non-negative definiteness of the four-dimensional "matrix"

$$||B_{jk}(t,t') - \frac{i\hbar}{2}[\chi_{jk}(t,t')-\chi_{kj}(t',t)]|| , \qquad (6.26)$$

which has one pair of discrete indices j, k, and one pair of continuous ones t, t'. This is the universal relation that connects the correlation functions for the fluctuations of an arbitrary linear system to the system's generalized susceptibilities.

6.5* The spectral representation

The spectral method is widely used in electronics and optics to simplify calculations and make them more transparent. This method is applicable to stationary systems, i.e. systems in which the response to an external action is independent of when the action occurs, and in which the statistical properties of the internal fluctuations do not change with time. For such a stationary system, the generalized susceptibilities and the correlation functions for the fluctuations depend only on the time difference $t-t'$ and not on absolute time. This permits one to represent these quantities by their Fourier transforms

$$\chi_{jk}(t,t') = \int_{-\infty}^{+\infty} \chi_{jk}(\omega)e^{i\omega(t-t')}\frac{d\omega}{2\pi} \; ,$$

$$B_{jk}(t,t') = \int_{-\infty}^{+\infty} S_{jk}(\omega)e^{i\omega(t-t')}\frac{d\omega}{2\pi} \; . \tag{6.27}$$

The Wiener-Khintchine theorem tells us that the $S_{jk}(\omega)$ appearing here are the spectral densities of the fluctuations.

By inserting equations (6.27) into the inequality (6.25) we obtain

$$\sum_{j,k=1}^{N}\int_{-\infty}^{+\infty} Q_j^*(\omega)Q_k(\omega)S_{jk}(\omega)\frac{d\omega}{2\pi}$$

$$\geq \frac{i\hbar}{2}\sum_{jk=1}^{N}\int_{-\infty}^{+\infty} Q_j^*(\omega)Q_k(\omega)[\chi_{jk}(\omega)-\chi_{kj}^*(\omega)]\frac{d\omega}{2\pi} \; . \tag{6.28}$$

Taking account of the symmetry relations

$$\chi_{jk}^*(\omega) = \chi_{jk}(-\omega) \; , \quad S_{jk}^*(\omega) = S_{jk}(\omega) = S_{kj}(\omega) \; , \tag{6.29}$$

which follow from the definition (6.27), it is easy to show that the following inequality is a necessary and sufficient condition for (6.28) to be satisfied:

$$\sum_{j,k=1}^{N} Q_j^*Q_k S_{jk}(\omega) \geq \frac{\hbar}{2}\left| i\sum_{j,k=1}^{N} Q_j^*Q_k[\chi_{jk}(\omega)-\chi_{kj}^*(\omega)]\right| \tag{6.30}$$

for all values of the frequency ω. Here $Q_1, \dots, Q_N$ are arbitrary complex numbers.

We shall now specialize these inequalities to the two most interesting cases: $N=1$ and $N=2$.

The case $N=1$ is that of the simplest linear system: one with a single input and no output, or, in the language of electronics, a system with just one pair of contacts (a two-pole system). Equation (6.30) says that the generalized susceptibility $\chi(\omega)$ of such a system and the spectral density $S(\omega)$ of its internal fluctuations are connecting by the following relation:

$$S(\omega) \geq \hbar |\,\text{Im}\chi(\omega)| \ . \tag{6.31}$$

This says that the source of the fluctuations is the dissipation, since the imaginary part of the generalized susceptibility (or in electronics language, the active part of the conductivity) describes the system's dissipation.

It is instructive to compare this inequality with the fluctuation-dissipation theorem:

$$S(\omega) = \hbar |\,\chi(\omega)\coth(\hbar\omega/2k_B T)| \ . \tag{6.32}$$

The right-hand sides of (6.31) and (6.32) coincide if the temperature T vanishes; in other words, the two formulae give the same lower limit for the fluctuations. Because the state of an equilibrium system is determined by a single parameter, its temperature T, the fluctuation-dissipation system (6.32) for such a system is able to express its level of fluctuations solely in terms of this parameter. Relation (6.31), by contrast, is valid for any linear system, including nonequilibrium ones for which the temperature is undefined. This forces the relation to take the form of an inequality: an absolute lower limit on the level of the fluctuations. Having an inequality rather than an equality is the price we must pay for the wider domain of applicability of (6.31) compared to (6.32).

The case $N = 2$ is that of a linear measuring device with one input and one output, i.e., the type of system on which this chapter focuses. In electronics such systems, with their two pairs of contacts (four-pole systems), are much more widely used than any other kind of linear system.

In view of the fact that relation (6.28) must be satisfied for all values of the arbitrary functions Q_1 and Q_2, a set of necessary and sufficient conditions for relation (6.30) to be satisfied is:

$$S_{jj} \geq \hbar |\,\text{Im}\chi_{jj}(\omega)| \ , \quad \text{for } j=1,2 \ , \tag{6.33}$$

$$[S_{11}(\omega)+\hbar\,\text{Im}\chi_{11}(\omega)][S_{22}(\omega)+\hbar\,\text{Im}\chi_{22}(\omega)] \geq |\,S_{21}+\frac{\hbar}{2i}[\chi_{21}(\omega)-\chi_{12}^*(\omega)]|^2 \ ,$$

$$\tag{6.34a}$$

$$[S_{11}(\omega)-\hbar\,\text{Im}\chi_{11}(\omega)][S_{22}(\omega)-\hbar\,\text{Im}\chi_{22}(\omega)] \geq |\,S_{21}-\frac{\hbar}{2i}[\chi_{21}(\omega)-\chi_{12}^*(\omega)]|^2 \ .$$

$$\tag{6.34b}$$

Strictly speaking, it is necessary to impose only one of the two inequalities (6.33) for one of the two values of j; the other will then be satisfied automatically by virtue of (6.34).

Condition (6.33) is the same as the inequality (6.31) for each of the system's two inputs/outputs separately. It describes the dissipation in the input and output. The inequalities (6.34), by contrast, reveal qualitatively new features of the system. These inequalities insist that the system have internal fluctuations even if the system is free of dissipation. More specifically: the absence of dissipation corresponds to a vanishing of the imaginary parts of the generalized susceptibilities

$$\text{Im}\chi_{jk}(\omega) \equiv 0 \ . \tag{6.35}$$

By inserting this condition into (6.34), we find that for a dissipation-free system the fluctuations must satisfy the inequality

$$S_{11}(\omega)S_{22}(\omega) - |S_{12}(\omega)|^2 \geq \frac{\hbar^2}{4}[\chi_{21}(\omega)-\chi_{12}(\omega)]^2$$

$$+ \hbar |[\chi_{21}(\omega)-\chi_{12}(\omega)]\text{Im}S_{12}(\omega)| \ . \tag{6.36}$$

It is easy to see that the minimum fluctuations are connected with an asymmetry between the forward (input to output) and reverse (output to input) transfer coefficients $\chi_{12}(\omega)$ and $\chi_{21}(\omega)$. For a symmetric system, which has $\chi_{12} \equiv \chi_{21}$, the inequality (6.36) degenerates into an identity, thereby permitting the system to be fluctuation-free. By contrast, for a measuring system with $\chi_{12}(\omega) = 0$ and $\chi_{21}(\omega) \neq 0$, relation (6.36) shows that fluctuations are unavoidable. It should be evident that these are the same fluctuations as accompany the process of continuous monitoring that was discussed in section 6.2.

6.6* Internal fluctuations of a linear measuring device

We now return to the linear measuring device that was analyzed in sections 6.1—6.3. We shall ignore the device's internal structure and treat it as a black box whose input accepts the measured observable $\hat{x}(t)$; see Fig. 6.2. The output signal $\tilde{x}(t)$ must be the sum of the input $\hat{x}(t)$ and some additional noise $\hat{x}_{\text{fluct}}(t)$ generated by the measuring device:

$$\tilde{x}(t) = \hat{x}(t) + \hat{x}_{\text{fluct}}(t) \ . \tag{6.37}$$

The measuring device also generates the fluctuating force $F_{\text{fluct}}(t)$ that acts back on the measured object.

For the type of measurement we are considering, nothing inserted into the output port can influence the output quantity $\tilde{x}(t)$. In other words, it must be possible to monitor the observable $\hat{x}$ in a QND way. This implies that the self-commutator of $\tilde{x}(t)$ must vanish precisely

$$[\tilde{x}(t),\tilde{x}(t')] \equiv 0 \ , \quad \text{for all } t \text{ and } t' \ . \tag{6.38}$$

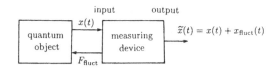

Fig. 6.2 General scheme for a continuous linear measurement.

This condition can be fulfilled, evidently, by virtue of the term $\hat{x}_{\text{fluct}}(t)$ in equation (6.37): the operator features of this quantity are compensating the operator features of the input observable $\hat{x}(t)$,

$$[\hat{x}_{\text{fluct}}(t),\hat{x}_{\text{fluct}}(t')] = -[\hat{x}(t),\hat{x}(t')] , \qquad (6.39)$$

thereby making the output signal $\tilde{x}(t)$, in effect, a c-number and not an operator.

In the language of the last section, the linear measuring device is a special case of a linear four-pole system with a special form of the matrix of generalized susceptibilities. In particular, the susceptibilities that describe the response to a signal applied to the output must vanish:

$$\chi_{12}(t,t') = \chi_{22}(t,t') = 0 . \qquad (6.40)$$

This permits us to rewrite the equations for the four-pole system in the form

$$\hat{Z}_1(t) = \int\limits_{-\infty}^{t} \chi_{11}(t,t')\hat{G}_1(t')dt' + \hat{Z}_1^0(t) ,$$

$$\hat{Z}_2(t) = \int\limits_{-\infty}^{t} \chi_{21}(t,t')\hat{G}_2(t')dt' + \hat{Z}_2^0(t) . \qquad (6.41)$$

Changing notation to

$$\hat{x}(t) = \hat{G}_1(t) , \quad \hat{F}(t) = \hat{Z}_1(t) , \quad \hat{F}_{\text{fluct}} = \hat{Z}_2^0(t) , \qquad (6.42)$$

and making the transition from the variable $\hat{Z}_2(t)$ to the same quantity referred to the system's input

$$\tilde{x}(t) = \int\limits_{-\infty}^{t} \chi_{21}^{-1}(t,t')\hat{Z}_2(t')dt' , \qquad (6.43)$$

where χ_{21}^{-1} is the inverse of χ_{21}

$$\int\limits_{t'}^{t} \chi_{21}^{-1}(t,t'')\chi_{21}(t'',t')dt'' = \delta(t-t') ,$$

we obtain

$$\hat{F}(t) = \int\limits_{-\infty}^{t} \chi_{11}(t,t')\hat{x}(t')dt' + \hat{F}_{\text{fluct}}(t) , \tag{6.44}$$

$$\tilde{x}(t) = \hat{x}(t) + \hat{x}_{\text{fluct}}(t) . \tag{6.45}$$

Here

$$\hat{x}_{\text{fluct}}(t) = \int\limits_{-\infty}^{t} \chi_{21}^{-1}(t,t')\hat{Z}_2^0(t')dt' \tag{6.46}$$

is the output noise of the measuring device, referred to the input.

Equation (6.44) describes the back action of the measuring device on the measured object. The first term has a regular character and is proportional to the measured quantity $\hat{x}(t)$. It is the dynamical back action, and it is permitted, in principle, to vanish (nothing prevents the measuring device from having $\chi_{11} = 0$). The second term is the fluctuational back action, and is fundamental and unavoidable. As we have seen earlier, it enforces the uncertainty principle for the measurement process.

By inserting equations (6.42)—(6.46) into equation (6.25), we find that the internal fluctuations of our linear measuring device must satisfy the following inequality:

$$\int\limits_{-\infty}^{+\infty} [Q_1^*(t)Q_1(t')B_F(t,t') + 2\text{Re}[Q_2^*(t)Q_1(t')B_{xF}(t,t')] +$$

$$+ Q_2^*(t)Q_2(t')B_x(t,t')]dtdt' + \hbar\text{Im}[\int\limits_{-\infty}^{+\infty} Q_2^*(t)Q_1(t)dt + \tag{6.47}$$

$$+ \int\limits_{-\infty}^{+\infty} Q_1^*(t)Q_1(t')\chi_{11}(t,t')dtdt'] \geq 0 .$$

Here $B_F(t,t')$ and $B_x(t,t')$ are the correlation functions for the input and output fluctuations, $B_{xF}(t,t')$ is their cross correlation function, and $Q_1(t)$ and $Q_2(t)$, as before, are arbitrary and in general complex functions.

The structure of equation (6.47) is substantially more complicated than that of the corresponding equation (6.17) of section 6.3, because it includes (i) a possible nonstationariness of the fluctuations, (ii) non-white spectral features and cross correlations of the fluctuations, and (iii) a nonzero input susceptibility $\chi_{11}(t,t')$ (in the language of electronics, a final input impedance for the measuring device). In various special cases of linear measuring devices, including those of sections 6.2 and 6.3, equation (6.47) takes on simpler forms. For example, if the meter has a vanishing input susceptibility and uncorrelated white noise,

$$B_F(t,t') = S_F \delta(t-t'), \quad B_x(t,t') = S_x \delta(t-t'), \quad B_{xF} = 0, \quad \chi_{11} = 0, \quad (6.48)$$

as was the case in section 6.2, (6.47) reduces to

$$\int_{-\infty}^{+\infty} \left[S_F |Q_1(t)|^2 + S_x |Q_2(t)|^2 + \hbar \mathrm{Im}[Q_2^*(t)Q_2(t)] \right] dt \geq 0. \quad (6.49)$$

It is not difficult to show that a necessary and sufficient condition for this relation to be satisfied, for all forms of the arbitrary functions $Q_{1,2}(t)$, is the inequality (6.7).

The most important special case of a linear measuring device is a stationary one, i.e. a device whose noises $x_{\mathrm{fluct}}(t)$ and F_{fluct} are stationary and are stationarily correlated, and whose input susceptibility χ_{11} [the only dynamical parameter entering into the relation (6.47)] depends only on the difference of its arguments t and t'. For such a device, it is natural to use a spectral description, as was done in the last section for general, stationary $2N$-poles. By transforming the inequality (6.47) in a manner analogous to that used in deriving equations (6.33) and (6.34), one can show that the spectral densities of the fluctuations in a classical (non-quantum) measuring device must obey the following inequalities:

$$S_F(\omega) \geq \hbar |\mathrm{Im}\chi_{11}(\omega)|, \quad (6.50)$$

$$S_x(\omega)S_F(\omega) - |S_{xF}(\omega)|^2 \geq \hbar |\mathrm{Im}[S_x(\omega)\chi_{11}(\omega) + S_{xF}(\omega)]| + \frac{\hbar^2}{4}. \quad (6.51)$$

Here $S_F(\omega)$ and $S_x(\omega)$ are the spectral densities of the input and output noise, $S_{xF}(\omega)$ is the spectral density associated with their cross correlation, and $\chi_{11}(\omega)$ is the Fourier transform of the input susceptibility of the measuring device.

VII   Nonlinear systems
for continuous measurements

7.1 Fluctuational and dynamical back action of the measuring device

During a continuous monitoring of a quantum object, the measuring device can act back on the object in general in two different ways: First, it can influence the evolution of the expectation value of the measured observables. This is called the "dynamical back action," and it is a regular, predictable effect. (An example is the dynamical back action produced by the dynamical rigidity of a capacity sensor, which will be discussed in chapter X.) Second, the measuring device can perturb the observables in a random way, increasing their uncertainties, i.e. producing random deviations from their expectation values. A key feature of linear, continuous measurements, as discussed in the last chapter, is that for them these two types of back action are completely independent. The fluctuational back action is fundamental and irremovable, while the dynamical back action can be avoided entirely by a suitable construction of the measuring device.

For nonlinear systems the situation is completely different, as one can readily see from a typical example of a nonlinear, continuous measurement: a monitoring of an oscillator's energy. In accord with the uncertainty relation, the energy measurement produces perturbations of the oscillator's phase, and as the measurement proceeds continuously, these perturbations cause the phase to diffuse in a Brownian-motion type way. Correspondingly, the oscillator's frequency will fluctuate around its initial value.

Suppose that, while the measurement is being made, a sinusoidal external force is acting on the oscillator, on resonance. The oscillator's response to the force will be the same as if the oscillator's frequency were fixed and the force's frequency were fluctuating (since the response depends only on the frequency difference). It is well known that the response to a force with fluctuating frequency is always less than that to a precisely harmonic force on resonance. In fact, an unsophisticated classical calculation shows that the magnitude of the response to the force decreases monotonically if the rate of the oscillator's phase diffusion is increased, i.e. if the precision of the energy monitoring is increased. Thus, the measuring device's *fluctuational* back action on the oscillator changes the oscillator's *dynamical* behavior.

The suppression of the response of a quantum object to an external force when the object's energy is being monitored continuously has been called the "watchdog effect".[33] A rigorous analysis of this effect (section 8.6) shows that the level of suppression is determined by the relative magnitudes of two timescales: the characteristic timescale τ_F for the force to change (more precisely the inverse width of its spectrum), and the time τ_o required to distinguish the oscillator's energy levels. The time τ_o is shorter the higher is the precision of the monitoring; completely precise monitoring corresponds to $\tau_o = 0$. If $\tau_o > \tau_F$, then the monitoring has almost no influence on the oscillator's response. If $\tau_o < \tau_F$, then the monitoring reduces the mean energy that the force maintains in the oscillator by a factor $\sqrt{\tau_o/\tau_F}$.

The watchdog effect is a particular case of the so-called "quantum Zeno paradox".[34] Recall the essence of Zeno's original paradox: a flying arrow, at each moment of time, is at a definite location in space, and thus it cannot move. By contrast with this purely logical paradox, the quantum Zeno paradox is a completely real effect. It shows up as a suppression of quantum transitions between discrete eigenstates of some observable, when the observable is being monitored continuously. (The observable must have a discrete spectrum of eigenvalues; it might, for example, be the energy of a harmonic oscillator or the angular momentum of some rotator.) This classically strange but quantum mechanically unavoidable effect was first observed experimentally in 1990.[35]

As should be clear from the above example, the suppression of the quantum transitions is not the measurement itself, i.e., not the experimenter's extraction of information about the measured quantity, but rather the back action of the measuring device on the measured object—a back action that is inevitable because of the uncertainty relations.

The analysis of the quantum Zeno paradox for general situations is very complicated, but it simplifies substantially if the monitoring has perfect precision. This case is analyzed in the next section, and then in the following sections of this chapter the measurements of arbitrary precision are treated. In particular, section 7.3 derives the evolution equations for nonlinear quantum systems that are being monitored continuously, and then section 7.4 discusses the quantum Zeno paradox for a two-level system. Later, in section 8.6, the results of this chapter will be used to analyze the watchdog effect for measuring devices that use oscillators as their quantum probes.

7.2 Quantum Zeno paradox for exact measurements

This section uses slightly more sophisticated mathematics than other unstarred sections. However, we recommend that all readers read it, so as to understand better the quantum Zeno paradox.

We presume that the observable q of some quantum system is being monitored continuously, and that the system begins in the nth eigenstate of q, with eigenfunction $\psi_n(x)$ (where x is the object's position or other generalized coordinate). As in the last chapter, we shall describe the continuous monitoring of q as the limit of a sequence of discrete, instantaneous measurements separated by small time intervals θ. Between measurements, i.e. during the intervals θ, the wave function's evolution is governed by the the Schrödinger equation

$$i\hbar\frac{\partial\psi}{\partial t} = \hat{H}\psi(x,t) \ . \tag{7.1}$$

Here $\hat{H}$ is the object's Hamiltonian, which, we recall, is nothing but the object's total energy, expressed in terms of its coordinate x and momentum p, but with p replaced by the momentum operator $(\hbar/i)(d/dx)$. We presume that this free evolution, if not interfered with by reduction of the wave function in any measurements, will cause the object, with some finite probability, to make a transition out of the eigenstate ψ_n of the observable q and into some other eigenstate, and we then ask ourselves how the monitoring of the object influences the probabilities for such transitions.

The first interval of free evolution begins at time $t = 0$, and for sufficiently small times t it produces the following wave function [note that this is the straightforward power-series solution of equation (7.1)]:

$$\psi(x,t) \simeq \left[1 + \frac{\hat{H}t}{i\hbar} + \frac{1}{2}\left[\frac{\hat{H}t}{i\hbar}\right]^2 + ... \right] \psi(x,0) \ .$$

Here $\psi(x,0) = \psi_n(x)$ is the initial value of the wave function. For our analysis it will be sufficient to keep terms only up through order t^2.

Immediately before the first measurement, i.e. at $t = 0$, this wave function is

$$\psi(x,0) \simeq \left[1 + \frac{\hat{H}0}{i\hbar} + \frac{1}{2}\left[\frac{\hat{H}0}{i\hbar} \right]^2 + ... \right] \psi_n(x) . \tag{7.2}$$

Correspondingly, the probability that this measurement will show the system to still be in the nth state (the probability that no transition will have been made) is

$$W_{nn} = \left| \int_{-\infty}^{+\infty} \psi_n^*(x)\psi(x,0)dx \right|^2$$

$$= \left| 1 + \frac{0}{i\hbar}H_{nn} - \frac{0^2}{2\hbar^2}(H^2)_{nn} \right|^2 \simeq 1 - \frac{0^2}{\hbar^2}[(H^2)_{nn} - (H_{nn})^2] , \tag{7.3}$$

where H_{nn} and $(H^2)_{nn}$ are the diagonal matrix elements of the operators $\hat{H}$ and $\hat{H}^2$:

$$\hat{H}_{nn} = \int_{-\infty}^{+\infty} \psi_n^*(x)\hat{H}\psi_n(x)dx ,$$

$$(\hat{H}^2)_{nn} = \int_{-\infty}^{+\infty} \psi_n^*(x)\hat{H}^2\psi_n(x)dx .$$

Because the quantity in square brackets in equation (7.3) is the variance $(\Delta E)^2$ of the object's energy in the initial state ψ_n, we can rewrite (7.3) as

$$W_{nn} = 1 - \frac{0^2}{\hbar^2}(\Delta E)^2 . \tag{7.4}$$

It should be evident that after k steps consisting of free evolution for a time 0 plus an instantaneous measurement of q, the probability that the object is still in its initial eigenstate ψ_n of q will be the kth power of expression (7.4):

$$W_{nn}(\tau) = [W_{nn}(0)]^k = \left[1 - \frac{0^2}{\hbar^2}(\Delta E)^2 \right]^{\tau/0} , \tag{7.5}$$

where $\tau = k0$ is the total time for these k steps, i.e. the total time of the sequence of measurements.

The limit of a continuous monitoring is achieved by taking the limit $0 \rightarrow 0$ in equation (7.5), with the total measurement time τ held fixed. When 0 has become sufficiently small, (7.5) tends to

$$W_{nn}(\tau) \rightarrow \exp\left[-\frac{\tau\theta}{\hbar^2}(\Delta E)^2\right],\qquad(7.6)$$

and in the ultimate limit of vanishing θ, it becomes

$$W_{nn}(\tau) = 1.\qquad(7.7)$$

In other words, the probability for the object to remain in its original eigenstate of q is precisely unity, no matter how large is the total measurement time τ. Correspondingly, the probability of a transition to some other eigenstate is zero.

Thus, when any observable of the object with a discrete spectrum of eigenvalues is monitored precisely, the object remains forever frozen in its initial state. It does not evolve at all.

7.3* The equation of motion for the density operator during a continuous monitoring

The purpose of this section is to develop a mathematical formalism for analyzing the evolution of nonlinear systems that are being monitored continuously. Since the formalism of chapter VI was based on the assumption of linearity, we cannot use it. In its place, we shall obtain an equation for the evolution of the quantum object's density operator during a continuous monitoring, and our equation will be valid for both nonlinear and linear systems.

We shall denote the object's Hamiltonian by $\hat{H}$ and its measured observable by $\hat{q}$; and, without producing much additional complexity in the analysis, we shall allow these operators to have explicit time dependences (in the Schrödinger picture).

In our derivation, we shall use a highly idealized model for the quantum probe: We shall take it to consist of a set of identical particles that do not interact with each other, and that have vanishing Hamiltonians. (That the Hamiltonian vanishes means, simply, that when the particle is decoupled from the rest of the world, its quantum state is independent of time, i.e. is conserved.) The jth particle interacts with the object during the time interval $(j-1)\theta \leq t \leq j\theta$, where θ is the duration of the interaction. (As above, the transition to continuous monitoring will be made by taking θ→0.) At the end of the jth interaction with the object, the observer performs a precise, direct measurement on the jth particle and from it infers information about the object's observable q. (The specific details of this idealized form of measuring device and measurement will drop out of the final evolution equation for the object's density operator.)

We shall presume, for simplicity, that the particles are prepared independently of each other, i.e. that their joint wave function can be decomposed into a product of one-dimensional wave functions for each of the particles independently. This assumption (which will imply in the continuum limit that the fluctuations of the measuring device have delta-function correlations, i.e. have white noise) constrains the generality of this analysis, compared to that of the last chapter. Despite this constraint, the analysis will reveal the most important of the physical effects that accompany continuous measurements of nonlinear systems.

Because the time θ is presumed arbitrarily short, the operators $\hat{\boldsymbol{H}}(t)$ and $\hat{q}(t)$ can be idealized as having constant values $\hat{\boldsymbol{H}}_j$ and $\hat{q}_j$ during the interaction with the jth particle. The full Hamiltonian during that interaction then has the form $\hat{\boldsymbol{H}}_j - \hat{q}_j \hat{y}_j$, where $\hat{y}_j$ is the generalized coordinate of the jth particle. The corresponding evolution operator from the beginning to the end of the jth interaction is

$$\hat{\boldsymbol{U}}_j = \exp\left[\frac{\hat{\boldsymbol{H}}_j - \hat{q}_j \hat{y}_j}{i\hbar}\theta\right] \simeq 1 + \frac{\theta}{i\hbar}(\hat{\boldsymbol{H}}_j - \hat{q}_j \hat{y}_j) - \frac{\theta^2}{2\hbar^2}(\hat{\boldsymbol{H}}_j - \hat{q}_j \hat{y}_j)^2. \quad (7.8)$$

We shall denote by $\hat{\rho}_j$ the object's density operator before the beginning of the jth interaction [i.e. at the time $(j-1)\theta$], and by $|\psi_j>$ the corresponding initial wave function of the jth particle (probe). Then, at the end of the jth interaction, the object will be in the state

$$\hat{\rho}_{j+1} = \mathrm{Tr}_j(\hat{\boldsymbol{U}}_j | \psi_j > \hat{\rho}_j < \psi_j | \hat{\boldsymbol{U}}_j^\dagger), \quad (7.9)$$

where Tr_j is the trace over the Hilbert space of the jth particle. This is the object's state if the experimenter does not look at the outcome of the measurement (i.e. does not collapse the coupled system onto that eigenstate of the jth particle which corresponds to the result of his exact, direct measurement of the particle) following the particle-object interaction. If, instead, the experimenter looks at the outcome of the measurement, then the state of the object will have a form analogous to equation (3.23).

By inserting the evolution operator (7.8) into expression (7.9), we obtain up through second order in θ

$$\hat{\rho}_{j+1} = \hat{\rho}_j + \frac{\theta}{i\hbar}[\hat{\boldsymbol{H}}_j - \hat{q}_j < \hat{y}_j >, \hat{\rho}_j] - \frac{\theta^2}{2\hbar^2}\Bigg\{ [\hat{\boldsymbol{H}}_j, [\hat{\boldsymbol{H}}_j, \hat{\rho}_j]]$$

$$- <\hat{y}_j>([\hat{\boldsymbol{H}}_j, [\hat{q}_j, \hat{\rho}_j]] + [\hat{q}_j, [\hat{\boldsymbol{H}}_j, \hat{\rho}_j]]) + <\hat{y}_j^2>[\hat{q}_j, [\hat{q}_j, \hat{\rho}_j]] \Bigg\}. \quad (7.10)$$

Here $<\hat{y}_j>$ and $<\hat{y}_j^2>$ are the mean and mean square coordinates of the particle j *before* its interaction with the object begins. We shall assume that $\langle \hat{y}_j \rangle = 0$; this corresponds (since the interaction Hamiltonian is proportional to $\hat{y}_j$) to assuming that the direct, regular back action of the quantum probe on the measured object vanishes. We shall also assume that in the transition $\theta \to 0$ the quantity

$$\sigma_F^2(t_j) \equiv \theta \langle \hat{y}_j^2 \rangle \tag{7.11}$$

remains constant; this will make the probe's fluctuational back action remain finite in the continuum limit. The quantity σ_F^2 is the variance of the probe's back action force on the quantum object, as one can see from the form $-\hat{q}_j \hat{y}_j$ of the interaction Hamiltonian.

From these two assumptions and equation (7.10), it follows that

$$\lim_{\theta \to 0} \frac{\hat{\rho}_{j-1} - \hat{\rho}_j}{\theta} = \frac{1}{i\hbar}[\hat{\boldsymbol{H}}_j, \hat{\rho}_j] - \frac{\sigma_F^2(t_j)}{2\hbar^2}[\hat{q}_j, [\hat{q}_j, \hat{\rho}_j]] . \tag{7.12}$$

By then taking the continuum limit we obtain our final form for the equation of motion of the object's density operator:

$$\frac{\partial \hat{\rho}(t)}{\partial t} = \frac{1}{i\hbar}[\hat{\boldsymbol{H}}(t), \hat{\rho}(t)] - \frac{\sigma_F^2(t)}{2\hbar^2}[\hat{q}(t), [\hat{q}(t), \hat{\rho}(t)]] . \tag{7.13}$$

This equation of motion differs from the standard one for a density operator by the inclusion of the new second term on the right-hand side. It should be evident on physical grounds that this new term should be inversely proportional to the precision of the monitoring (the more precise the monitoring, the larger should be this term). We shall now derive a quantitative relation connecting the monitoring with the quantity σ_F^2.

We shall denote by P_j the momentum of particle j, which is canonically conjugate to its coordinate y_j:

$$[\hat{y}_j, \hat{P}_j] = i\hbar . \tag{7.14}$$

The interaction with the object produces, in the Heisenberg picture, the following change in this momentum:

$$\hat{P}_j(\theta) = \hat{\boldsymbol{U}}_j^{\dagger} \hat{P}_j \hat{\boldsymbol{U}}_j = \hat{P}_j + \hat{q}_{j+\frac{1}{2}} \cdot \theta , \tag{7.15}$$

where

$$\hat{q}_{j+\frac{1}{2}} = \hat{q}_j + \frac{\theta}{2i\hbar}[\hat{q}_j, \hat{\boldsymbol{H}}_j] \tag{7.16}$$

is the value of the operator $\hat{q}_j$ at the midpoint of the interaction interval, and $\hat{\boldsymbol{U}}_j$ is the evolution operator (7.8). The output signal of the jth probe particle is $\hat{P}_j(\theta)$. This output signal is measured precisely and directly in the second stage of the experiment, but as was discussed above, in evolving the object's density operator, we presume that the experimenter throws away the result of that measurement. This output signal can be regarded equally well, by virtue of equation (7.15), as

$$\tilde{q}_{j+\frac{1}{2}} \equiv \frac{\hat{P}_j(\theta)}{\theta} = \hat{q}_{j+\frac{1}{2}} + \frac{\hat{P}_j}{\theta} .$$ (7.17)

This readout quantity is a superposition of the object's observable $q_{j+\frac{1}{2}}$ and the fluctuations (in P_j) that the probe particle had before the measurement began. The strength of the fluctuations is described by the correlation matrix of the last chapter, which in this case is diagonal:

$$B_{jj'} = \frac{<\hat{P}_j^2>}{\theta} \delta_{jj'} .$$ (7.18)

(We presume that $<\hat{P}_j> = 0$; i.e., the measuring device does not introduce any bias into the measured result.)

The variance $<\hat{P}_j^2>$ of the momentum is related to the variance $<\hat{y}_j^2>$ of the coordinate by the uncertainty relation

$$<\hat{y}_j^2><\hat{P}_j^2> - |<\hat{y}_j \circ \hat{P}_j>|^2 \geq \frac{\hbar^2}{4}$$ (7.19)

(where we recall that the means of $\hat{y}_j$ and $\hat{P}_j$ vanish), which follows from equation (7.14) and the fact that all the particles are independent. In the continuum limit, $\theta \to 0$, the form of this relation clearly does not change. By taking that limit holding fixed

$$\sigma_q^2(t) = \frac{<\hat{P}_j^2>}{\theta} ,$$ (7.20)

$$\sigma_{qF}^2(t_j) = |<\hat{y}_j \circ \hat{P}_j>|^2 ,$$ (7.21)

and the $\sigma_F^2(t_j)$ of (7.11), we bring equation (7.19) into the form

$$\sigma_F^2(t) \cdot \sigma_q^2(t) - \sigma_{qF}^2(t) \geq \frac{\hbar^2}{4} .$$ (7.22)

According to equation (7.18), in the continuum limit, the quantity $\sigma_q(t)$ determines the correlation function of the noise that the measuring device adds to the measured quantity:

$$B_q(t,t') = \sigma_q^2 \delta(t-t') .$$ (7.23)

Equation (7.22) exhibits the sense in which the strength of the influence of the monitoring on the object's state (the last term in equation (7.13), which is proportional to σ_F^2) is inversely proportional to the accuracy σ_q^2 of the monitoring.

7.4* Quantum Zeno paradox for approximate measurements

The standard example of the quantum Zeno paradox is the evolution of a two-level system when the level in which the system sits is continuously monitored. One can calculate the evolution of this simple system exactly and see how it depends on the accuracy of the monitoring. We shall do so, making only one simplifying assumption: that the accuracy of the monitoring is constant, i.e. that the quantities σ_F, σ_q, and σ_{qF} of equation (7.22) are independent of time.

The general, free evolution of such a system consists of periodic oscillations between its two levels. Denoting these level's eigenstates by $|+>$ and $|->$, we can write the action of the system's free evolution operator $\hat{U}(t)$ in the form

$$\hat{U}(t)|+> = \cos\frac{\Omega t}{2}|+> + i\sin\frac{\Omega t}{2}|->,$$

$$\hat{U}(t)|-> = i\sin\frac{\Omega t}{2}|+> + \cos\frac{\Omega t}{2}|->. \tag{7.24}$$

The parameter Ω is the frequency of oscillation of the transition probability

$$|<+|\hat{U}(t)|->|^2 = \frac{1}{2}(1-\cos\Omega t). \tag{7.25}$$

Let us denote by $\hat{I}$ the unit 2×2 matrix, and by $\hat{S}_x$, $\hat{S}_y$, $\hat{S}_z$ the Pauli spin matrices. In this notation the evolution operator $\hat{U}(t)$ takes the form

$$\hat{U}(t) = \hat{I}\cos\frac{\Omega t}{2} + i\hat{S}_x\sin\frac{\Omega t}{2}. \tag{7.26}$$

It is easy to see that the Hamiltonian corresponding to this evolution operator is

$$\hat{H} = -\frac{\hbar\Omega}{2}\hat{S}_x. \tag{7.27}$$

A familiar example of such a system is a spin-½ particle in a magnetic field. If the field is along the x-axis, and $|+>$ and $|->$ are the states with fixed projections of the spin along the z-axis, then the Hamiltonian of the spin degree of freedom will have the form (7.27). The frequency Ω in this case is twice the spin-precession frequency: $\Omega = 2\mu H/\hbar$ where H is the magnetic field strength and μ is the particle's magnetic moment.

A second example is a pair of identical, coupled oscillators which together contain precisely one quantum of excitation. If the coupling is sufficiently weak, then in the rotating-wave approximation the Hamiltonian takes the following form:

$$\hat{H} = \hbar\omega(\hat{a}_1^\dagger\hat{a}_1 + \hat{a}_2^\dagger\hat{a}_2 + 1) - \frac{\hbar\Omega}{2}(\hat{a}_1^\dagger\hat{a}_2 + \hat{a}_2^\dagger\hat{a}_1) \ ,$$

where $\hat{a}_{1,2}^\dagger$ and $\hat{a}_{1,2}$ are the annihilation and creation operators for the first and second oscillators, ω is the frequency of the individual oscillators, Ω is the coupling frequency, and the assumption of weak coupling has the explicit form $\Omega \ll \omega$. The two states corresponding to localization of the quantum in one or the other oscillators create a basis for the Hilbert space of all one-quantum states. The Hamiltonian $\hat{H}$ in this basis has the 2×2 matrix form

$$\hat{H} = 2\hbar\omega\hat{I} - \frac{\hbar\Omega}{2}\hat{S}_x \ .$$

The first term can be omitted because it produces only a phase factor $e^{-2i\omega t}$ in the evolution operator, and such a phase factor has no physical consequences. The remaining second term coincides with the Hamiltonian (7.27).

For either of these examples, or any other two-state system, we shall now examine the effect of coupling to a measuring device that monitors continuously which of the two states, $|+>$ or $|->$, the system is in. For such a monitoring, the eigenstates of the measured observable are $|+>$ and $|->$, and its corresponding eigenvalues, without loss of generality, can be set equal to +1 and −1. Then, for the example of a spin-½ particle, the output signal will be +1 if the spin is along the z-axis, and −1 if it is opposite; and for the coupled-oscillator system, the output will be +1 or −1 depending on which oscillator the quantum is in. The measured observable, with these eigenstates and eigenvalues, has the simple form

$$\hat{q} = \hat{S}_z \ . \tag{7.28}$$

In the representation generated by $\hat{q}$, the density operator is a 2×2 matrix; and taking account of its normalization, $\text{Tr}\hat{\rho} = 1$, it can be written in the form

$$\hat{\rho} = \frac{1}{2}(\hat{I} + \rho_x\hat{S}_x + \rho_y\hat{S}_y + \rho_z\hat{S}_z) \ . \tag{7.29}$$

By inserting expressions (7.27)—(7.29) into the evolution equation (7.13) for the density operator, we obtain

$$\frac{\partial}{\partial t}(\rho_x\hat{S}_x + \rho_y\hat{S}_y + \rho_z\hat{S}_z) = \Omega(-\rho_y\hat{S}_z + \rho_z\hat{S}_y) - \frac{2}{\tau_o}(\rho_x\hat{S}_x + \rho_y\hat{S}_y) \ , \tag{7.30}$$

where

$$\tau_o = \frac{\hbar^2}{\sigma_F^2} \leq \frac{4\sigma_q^2}{1 + (2\sigma_{qF}/\hbar)^2} \tag{7.31}$$

[cf. equation (7.22)]. To clarify the physical meaning of the parameter τ_o, notice that the variance of the output signal in the optimal case, when $\sigma_{qF} = 0$ and there is an equality in (7.22), is $(\Delta\tilde{q})^2 = \tau_o/4\tau$, where τ is the averaging time.

From equation (7.30) we obtain equations of motion for the components of the density operator:

$$\frac{\partial\rho_x}{\partial t} + \frac{2}{\tau_o}\rho_x = 0 \,, \quad \frac{\partial\rho_y}{\partial t} + \frac{2}{\tau_o}\rho_y - \Omega\rho_z = 0 \,, \quad \frac{\partial\rho_z}{\partial t} + \Omega\rho_y = 0 \,. \quad (7.32)$$

Now, let us suppose that the measured object initially is localized in one of its two states, say $|+\rangle$, i.e. initially $\rho(0) = |+\rangle\langle+|$. The corresponding components of the initial density matrix, as inferred from equation (7.29), are

$$\rho_x(0) = \rho_y(0) = 0 \,, \quad \rho_z(0) = 1 \,. \quad (7.33)$$

The solution of the equations of motion (7.32) for this initial state has the following form:

$$\rho_x(t) = 0 \,,$$

$$\rho_z(t) = \begin{cases} (\cos\Omega't + \dfrac{1}{\Omega'\tau_o}\sin\Omega't)e^{-t/\tau_o} & \text{if } \Omega\tau_o > 1 \,, \\[2mm] \dfrac{1}{\tau'-\tau''}(\tau'e^{-t/\tau'} - \tau''e^{-t/\tau''}) \,, & \text{if } \Omega\tau_o < 1 \,, \end{cases} \quad (7.34)$$

where

$$\Omega' = \sqrt{\Omega^2 - \tau_o^{-2}} \,, \quad \tau'^{-1} = \tau_o^{-1} - \sqrt{\tau_o^{-2} - \Omega^2} \,, \quad \tau''^{-1} = \tau_o^{-1} + \sqrt{\tau_o^{-2} - \Omega^2} \,.$$

Now, equation (7.29) implies that the probabilities of finding the object in the states $|+\rangle$ and $|-\rangle$ are

$$W_\pm = \frac{1 \pm \rho_z(t)}{2} \,. \quad (7.35)$$

From equation (7.34) one can see that, as $t \to \infty$, $\rho_z \to 0$. This means that, when the experimenter never looks at the output of the measuring device, the device's back action on the measured object drives it asymptotically into an equilibrium state in which the probabilities for the two levels are equal. If $\Omega\tau_o > 1$, then the evolution toward equilibrium is accompanied by oscillations of the probabilities $W_\pm$. These oscillations are analogous to the ones that occur in the object's free evolution, i.e. when no measurement is being made. As the monitoring accuracy is increased (i.e. as τ_o

is decreased), the frequencies of these oscillations decreases until, for $\Omega\tau_o = 1$, the oscillations are completely suppressed, and the probabilities $W_\pm$ asymptote to ½ monotonically.

In the ultimate case of very high-accuracy monitoring, $\Omega\tau_o \ll 1$, the function $\rho_z(t)$ for times that are not too small can be approximated by the exponential

$$\rho_z(t) \simeq e^{-t/\tau_1} , \tag{7.36}$$

where

$$\tau_1 = \frac{2}{\Omega^2 \tau_o} \tag{7.37}$$

is the mean lifetime of the object in its initial level.

Thus, the measuring device changes the evolution of the object substantially if the parameter τ_o, which characterizes the monitoring accuracy, is less than the characteristic timescale Ω^{-1} for free evolution. In this case, the object's lifetime in its initial state is inversely proportional to τ_o. In the limit $\tau_o \to 0$, i.e. when the monitoring is exact, the object is frozen forever in its initial state, and it never evolves.

We note, in conclusion, that these results imply the intriguing conclusion that for any choice of the total length of a continuous monitoring, one can choose a sufficiently high accuracy of monitoring (sufficiently small τ_o) as to completely suppress the oscillations of the measuring device's output.

VIII   Detection of classical forces

8.1 Aspects of quantum limits for the detection of a classical force

The development of the theory of QND measurements was triggered by the problem of detecting a small classical force that acts on a quantum probe—e.g. on a free mass or an oscillator; see the review article[36] and the monograph[29]. This problem arose most especially in gravitational-wave experiments. Astrophysical estimates indicate that the gravitational waves from cosmic sources should be so strong that they behave almost perfectly classically. During the measuring time the number of gravitons that pass within one wavelength of the measuring device is enormously larger than unity; and correspondingly, as would be the case if the waves were electromagnetic, quantum effects in the radiation are utterly negligible.

A central feature of gravitational waves is their extremely weak interaction with matter. Because of this weakness, the response of the measuring device's probe object to the waves is so small that it may be comparable to the object's quantum mechanical uncertainties. At the same time, the weakness of the interaction guarantees that there is almost no back action of the probe on the gravitational-wave field. As a result, the gravitational wave acts on the probe as through it were a precisely classical force (i.e. a force that is independent of the probe's quantum state).

It is easy to give other examples of classical forces that act on quantum probes. One example arises in the measurement of very small electric charges or magnetic moments of a rather massive particle using a quantum probe. If the particle's mass is large enough, then one can ignore the back action of the probe on the particle's trajectory and speed, and correspondingly, one can regard the particle as exerting a purely classical force on the probe.

In any such case of a classical force and a quantum probe, any simple-minded attempt to apply the uncertainty relations gives a trivial result: there is no limit on the smallness of the force that, in principle, can be detected. The origin of this is simple: there is no restriction, in principle, on the precision with which a classical observable can be monitored. Nevertheless, because there is a quantum probe between the classical force and the classical observer, there in fact can be limitations on the sensitivity. Such limitations can arise in two ways:

First, limitations can be produced by the insertion of excess energy into the probe during the measurement. More specifically: There is a universal relation between the measurement sensitivity and the amount of energy that the classical parts of the measuring device insert into the quantum probe. This relation comes about as follows. The measurement of any probe observable must be accompanied, according to the uncertainty relations, by perturbations of other probe observables (e.g., the measurement of the probe's position must be accompanied by a perturbation of its momentum). When exerting its perturbation, the remainder of the measuring device will inevitably insert energy into the probe. The higher is the measurement precision, the stronger will be the perturbation and the larger will be the inserted energy. When the inserted energy is large enough, it produces effects in the probe (e.g. electrical breakdown, mechanical destruction, or simply driving the probe beyond its linear dynamical range) that limit its precision of measurement. The level of these limitations is different for different kinds of probes, but like the standard quantum limits of section 1.4, these limitations depend only on fundamental physical constants (e.g. Planck's constant and the charges and masses of elementary particles). The next chapter is dedicated to the class of limitations on measurement precision.

Second, practical difficulties may prevent the experimenter from using an optimal procedure of measurement (one that, in principle, would permit unlimited sensitivity); and, being forced back to some "standard," nonoptimal procedure, the experimenter may encounter unavoidable precision limitations. The standard procedures are generally based on a continuous monitoring of some coordinate of the probe, with a time-independent monitoring precision. Examples, when the coordinate monitored is the position of some probe mass, include optical interferometer sensors, capacity sensors with constant pumping, and sensors based on SQUIDs (superconducting quantum interference devices). For these standard schemes, the precision is constrained by the standard quantum limits of section 1.4.

As an example of this second type of limitation, suppose that one wants to detect a classical, external force that acts on a probe mass, and one has been forced by practical limitations to monitor the mass's position rather than its momentum. To detect the force, the measuring device must give information about the probe's position at two or more different moments of time. The measurement of position at the earlier moments of time perturbs the particle's momentum, and these perturbations produce uncertainties in the position at the later moments of time, thereby imposing the standard quantum limit on the overall measurement accuracy; see section 1.4. If the external force is F and it lasts for a time τ_F, then the particle's momentum changes between measurements by $F \cdot \tau_F$, and to detect this change, the force must be at least as strong as

$$F \geq \frac{\Delta P_{SQL}}{\tau_F} = \frac{1}{\tau_F} \sqrt{\frac{\hbar m}{2\tau}} \ , \qquad (8.1)$$

where ΔP_{SQL} is the standard quantum limit for momentum [equation (1.22)] and τ is the measurement time. In gravitational-wave experiments, the time averaged force vanishes, so the wave leaves the mass with the same final momentum as it began with. Correspondingly, in this case τ must be shorter than the wave's duration τ_F, and the maximum sensitivity (for a continuous monitoring of the probe mass's position) is given by the following standard quantum limit for the mass's sensitivity to the classical force:

$$F_{SQL} = \sqrt{\frac{\hbar m}{2\tau_F^3}} \ . \qquad (8.2)$$

Another important example of a quantum probe is a harmonic oscillator. The motion of a free oscillator can be written in the form

$$x(t) = X_1 \cos\omega t + X_2 \sin\omega t \ ,$$

where X_1 and X_2 are the oscillator's quadrature amplitudes, which are related to its initial position and momentum by

$$X_1 = x(0) \ , \qquad X_2 = \frac{P(0)}{m\omega} \ .$$

Here m and ω are the oscillator's mass and frequency. The uncertainty relations for the quadrature amplitudes (corresponding to $\Delta x \cdot \Delta P \geq \hbar/2$ for the initial position and momentum), have the following form

$$\Delta X_1 \cdot \Delta X_2 \geq \frac{\hbar}{2m\omega} . \tag{8.3}$$

When the position is monitored continuously, symmetry implies that both quadrature amplitudes are measured with the same precision:

$$\Delta X_1 = \Delta X_2 \geq \sqrt{\frac{\hbar}{2m\omega}} .$$

The standard quantum limit (1.24) for the oscillator's position follows from this.

An external force acting on the oscillator changes its position by

$$\delta x = \xi \frac{F \tau_F}{m\omega} , \tag{8.4}$$

where F is the amplitude of the force, τ_F is its duration, and ξ is a factor that depends on the form of the force (e.g., for $\omega\tau_F \ll 1$, i.e. a short kick, $\xi = 1$; and for a force that oscillates sinusoidally on resonance with the oscillator, $\xi = \frac{1}{2}$). The force can be detected if

$$\delta x > \sqrt{2}\,\Delta x_{SQL} .$$

The factor $\sqrt{2}$ takes account of the necessity for at least two measurements of the oscillator: one before and one after the force acts. From this limit we infer that the force can be detected only if

$$F \geq F_{SQL} = \frac{1}{\xi\tau_F}\sqrt{\hbar\omega m} . \tag{8.5}$$

This is the so-called standard quantum limit for an oscillator's sensitivity to the classical force. (For a more rigorous analysis, see section 8.4.)

Dissipation in the probe can also place important limits on sensitivity. Rigorously speaking, the limits due to dissipation are not quantum in origin. However, they define a level of sensitivity (called the probe's "potential sensitivity") that cannot be surpassed by any method, no matter how sophisticated.

According to the classical Nyquist formula, accompanying any dissipation in the probe there is a random force that acts on the probe. The spectral density of this force is

$$S_{thermal} = 2k_B TH$$

where T is the temperature and H is the coefficient of friction. To be detectable in the presence of this fluctuating force, an external force must have amplitude

$$F \geq \frac{1}{\xi}\sqrt{\frac{S_{thermal}}{\tau_F}} = \frac{1}{\xi}\sqrt{\frac{2k_B TH}{\tau_F}} .$$

If the temperature is small enough, then in place of the Nyquist formula we must use the corresponding Callen-Velton quantum formula

$$S_{\text{thermal}} = \hbar\omega H \coth\frac{\hbar\omega}{2k_B T} \; ,$$

where ω is the frequency at which one observes the force. In the quantum limit, $k_B T \ll \hbar\omega$, this spectral density becomes

$$S_{\text{thermal}} = \hbar\omega H \; ,$$

and the expression for the potential sensitivity becomes

$$F \geq \frac{1}{\xi}\sqrt{\frac{\hbar\omega H}{\tau_F}} \; .$$

8.2 Quantum probe oscillator

The principal underlying methods to beat the standard quantum limit were formulated in chapter IV: the quantum probe must avoid entanglement with the uncertainty relations. When measuring a quantum object, this can be achieved in only one way: design the probe so it "sees" only the measured observable. By contrast, when measuring a classical force, entanglement in the uncertainty principle can be avoided by either of two methods: *First*, design the probe so it passes on to the rest of the measuring device only information about the measured force, and no information about the probe's quantum state. In this case, the classical nature of the force permits it to be monitored with very high precision. *Second*, perform a direct quantum nondemolition (QND) measurement on the probe; i.e. measure a set of probe observables that respond to the external force, but do not get perturbed in the process of their measurement. We shall illustrate these two methods using a simple example of a probe: a harmonic oscillator.

For a probe harmonic oscillator, one can implement the first method by measuring the average value of the oscillator's coordinate during one oscillation eigenperiod.[37] When no external force acts, this averaged coordinate is identically zero; and therefore the part of the measuring device that follows the probe, and the observer beyond it, receive no information at all about the oscillator's state. If an external force acts, then the averaged coordinate will be nonzero, but it will still carry no information about the oscillator's state. Correspondingly, if the precision of measurement of the averaged coordinate is high enough, the force can be detected.

Let us analyze a possible method of realizing this measurement procedure when the oscillator is an *LC* circuit and the classical force is a small e.m.f. acting on the circuit. (Recall that an *LC* circuit is an electrical analog of a mechanical oscillator: the inductance *L* plays the role of the mass, the inverse capacity $1/C$ is the rigidity, the charge on the capacitor is the coordinate, and e.m.f. is the external force.) Let us use as the quantum probe an electron that flies through the capacitor; see section 3.2. One can show that, if the time the electron takes to traverse the capacitor is equal to one period of the electrical oscillations, then the deflection of the electron does not depend on the circuit's initial state, but instead depends only on the time integral of the e.m.f. over the flight. The circuit's state is perturbed by the passing electron, but the state returns to its original value at the end of the traversal. The minimum measurable e.m.f. in this case is given by

$$V = \frac{1}{\xi \tau_V} \frac{d}{e} \Delta P \ ,$$

where τ_V is the duration of the e.m.f., d is the distance between the capacitor plates, e is the charge of the electron, ΔP is the initial uncertainty in the electron's transverse momentum, and the parameter ξ has the same meaning as in equation (8.5). The value of ΔP is related to the uncertainty Δy in the transverse position of the electron by

$$\Delta y \cdot \Delta P \geq \frac{\hbar}{2} \ .$$

In turn, the position uncertainty Δy is limited by the gap between the plates: $\Delta y \leq d/2$. From this it follows that the minimum detectable e.m.f. is

$$V \geq \frac{1}{\xi \tau_V} \frac{\hbar}{e} \ .$$

We turn attention next to an example of the second method of avoiding entanglement with the uncertainty principle when measuring a classical force: a stroboscopic measurement of the coordinate of a harmonic oscillator.[38] The two-time uncertainty relation for the coordinate has the following form:

$$\Delta x(t) \cdot \Delta x(t') \geq \frac{\hbar}{2m\omega} |\sin\omega(t - t')| \ . \tag{8.6}$$

The right-hand side is zero if

$$t - t' = \frac{n\pi}{\omega} \ ,$$

where n is any integer. Thus, if the two times are separated by an integral number of half periods, a measurement of the coordinate at time t

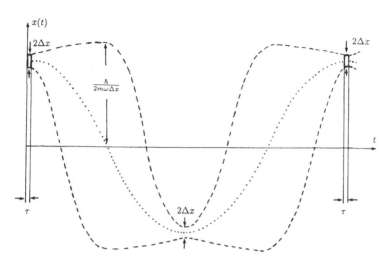

Fig. 8.1 The evolution of an oscillator when a stroboscopic measurement is performed.

need not perturb the value of the coordinate at time t'.

If the measurement precision is Δx, then the momentum will be perturbed by the amount $\hbar/2\Delta x$, and this will cause a spread of the oscillator's wave function that reaches a maximum at the time $t+\pi/2\omega$; see Fig. 8.1. The width of the wave function at this moment is $\hbar/2m\,\omega\Delta x$. After this moment, the wave function becomes narrower and narrower, until at $t = \pi/\omega$ it has squeezed itself back down to its original width Δx. This spreading and narrowing is then repeated periodically. By making a second coordinate measurement at the moment $t+\pi/\omega$, the observer can determine whether an external force has acted on the oscillator between measurements. If there was such a force and (for simplicity) it was impulsive, then the coordinate at the moment of the second measurement will be

$$x(t+\pi/\omega) = -x(t) + \frac{F\,\tau_F}{m\,\omega}\sin\omega(t_F-t) \ ,$$

where t_F is the time at which the force acted. Thus, for a favorable value of t_F ($t_F = t+\pi/2\omega$), it is possible to detect a force as small as

$$F = \frac{m\,\omega\Delta x}{\tau_F} \ . \tag{8.7}$$

Here, as in the previous example, the sensitivity is not limited by the standard quantum limit: the smaller one makes Δx, the higher is the sensitivity. However, one must pay attention to the important fact that the higher is the precision with which x is measured, the larger is the perturbation of the oscillator's momentum, and thus the larger is the amplitude of the resulting fluctuations in the oscillator's amplitude:

$$A_{fluct} = \frac{\hbar}{2m\,\omega\Delta x}\,.$$

The energy of these fluctuations

$$E_{fluct} = \frac{m\,\omega^2 A_{fluct}^2}{2} = \hbar\omega(n_{fluct}+1/2) \tag{8.8}$$

is inserted into the probe oscillator by the rest of the measuring device. By combining these relations, we obtain

$$\Delta x = \frac{\hbar}{2m\,\omega A_{fluct}} = \frac{1}{2}\sqrt{\frac{\hbar}{m\,\omega(2n_{fluct}+1)}}\,,$$

and by then inserting this into equation (8.7), we arrive at the following connection between the minimum detectable force and the mean number n_{fluct} of quanta of excitation that are inserted into the oscillator during the measurement:

$$F = \frac{1}{2\tau_F}\sqrt{\frac{\hbar\omega m}{2n_{fluct}+1}} = \frac{F_{SQL}}{2\sqrt{2n_{fluct}+1}}\,. \tag{8.9}$$

Since there is a limit on the amount of energy that any real oscillator can handle, and still function as an accurate probe, the sensitivity cannot be infinitely high.

8.3 Continuous quantum nondemolition monitoring

In the design of a real device for measuring a weak force, it is convenient to operate in a "stationary regime," i.e. to monitor the chosen probe observable with a time-independent sensitivity. In section 4.2 it was shown that to avoid the standard quantum limit in such a case, the probe observable must be an integral of the motion. Continuous quantum nondemolition monitoring of an integral of the motion, as a method of detecting an external force that acts on the probe, is called "QND readout of the force."

For a harmonic oscillator the integrals of motion are the two quadrature amplitudes X_1, X_2, and the energy. Let us analyze the QND readout of a force that is based on continuous monitoring of one of the quadrature amplitudes.[39] The quadrature amplitudes can be expressed in terms of the instantaneous values of the oscillator's coordinate and momentum as follows:

$$X_1 = x(t)\cos\omega t - \frac{P(t)}{m\omega}\sin\omega t \; ,$$

$$X_2 = x(t)\sin\omega t + \frac{P(t)}{m\omega}\cos\omega t \; . \qquad (8.10)$$

By inserting into these formulae the expressions for $x(t)$ and $P(t)$ in terms of the external force $F(t)$ and their initial values, we obtain

$$X_1 = x(0) - \frac{1}{m\omega}\int_{-\infty}^{t} F(t')\sin\omega t' \, dt' \; ,$$

$$X_2 = \frac{P(0)}{m\omega} + \frac{1}{m\omega}\int_{-\infty}^{t} F(t')\cos\omega t' \, dt' \; . \qquad (8.11)$$

By measuring either X_1 or X_2, one can detect either the sine or the cosine component of the external force. The higher is the precision of the measurement, the more strongly will the other quadrature amplitude be perturbed.

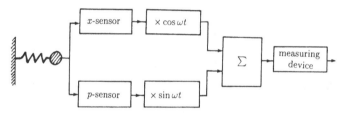

Fig. 8.2 A scheme for measuring one quadrature amplitude of an oscillator.

The conceptual method to monitor a quadrature amplitude is rather sophisticated; see Fig. 8.2: It entails two sensors, one for the oscillator's coordinate and the other for its momentum, two devices to multiply their output signals by $\cos\omega t$ and $\sin\omega t$, and a device to add the two resulting signals. All this must be done at the quantum level; i.e., it must be part of the quantum probe (see chapter III).

If one is willing to confine one's monitoring to a narrow bandwidth $\Delta\omega \ll \omega$ near the oscillator's eigenfrequency ω (and in doing so, use a long measuring time $\tau_{\text{measure}} \gg 1/\omega$), then one can do away with the momentum sensor—a device that in practice is especially sophisticated to construct. See Fig. 8.3. In this case, the output of the coordinate sensor, after multiplication by $\cos\omega t$ (at point A in Fig. 8.3) has the form

$$x(t)\cos\omega t = X_1\cos^2\omega t + X_2\cos\omega t \cdot \sin\omega t \; . \qquad (8.12)$$

When this is averaged over time, the second term vanishes and the first

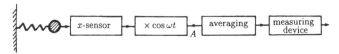

Fig. 8.3 Scheme for a quasi-QND measurement of an oscillator's quadrature amplitude.

becomes $\frac{1}{2}X_1$. Thus, this procedure, like the preceding one, measures only one quadrature amplitude and correspondingly only one quadrature component of the force. A concrete scheme for a measurement of this type will be analyzed in chapter X.

We next turn attention to the QND readout of a classical force based on a continuous monitoring of an oscillator's energy. (For a complete analysis, see section 8.6.) The force changes the amplitude A of the oscillator's motions by an amount

$$\delta A = \xi \frac{F\tau_F}{m\omega}\sin\phi \,,$$

where the value of ϕ is determined by the oscillator's initial phase. In a quantum state with well-defined energy, the phase is spread uniformly over the range from 0 to 2π, and correspondingly one cannot predict whether the force will increase the amplitude or decrease it. The rms change of the oscillator's energy in this case is

$$\delta E = m\omega^2 A \sqrt{<(\delta A)^2>} = \xi\frac{AF\omega\tau_F}{\sqrt{2}} \,.$$

The force can be detected only if this change is at least one quantum,

$$\delta E \gtrsim \hbar\omega \,.$$

Thus, the minimum detectable force is

$$F \geq \frac{1}{\xi\tau_F}\frac{\sqrt{2}}{\omega A}\delta E = \frac{1}{\xi\tau_F}\frac{\sqrt{2}\cdot\hbar}{A} \,,$$

or because

$$A = \sqrt{\frac{2E}{m\omega^2}} = \sqrt{\frac{2\hbar(n+\frac{1}{2})}{m\omega}} \,,$$

where n is the number of quanta in the oscillator,

$$F \geq \frac{1}{\xi\tau_F}\sqrt{\frac{\hbar\omega m}{n+\frac{1}{2}}} = \frac{F_{SQL}}{\sqrt{n+\frac{1}{2}}} \,. \tag{8.13}$$

The main conclusions of the first three sections of this chapter can be summarized by two statements:

1. If one uses standard methods to extract information from the probe (i.e., one monitors continuously one of the probe's coordinates), the measurement's sensitivity to an external, classical force is constrained by the standard quantum limit [equation (8.2) for a free mass, or (8.5) for an oscillator].

2. By using a QND monitoring to extract the information, one can achieve higher sensitivity, but there are still limitations connected with thermal dissipation (the "potential sensitivity") and with the amount of energy that the probe can deal with and still function properly (see the next chapter).

8.4* Standard quantum limit for an oscillator

In this section we shall give a rigorous derivation of the exact form of the standard quantum limit for the sensitivity with which one can measure a classical force acting on a probe oscillator. The system of measurement is shown schematically in Fig. 8.4. The classical external force $F(t)$ drives the probe oscillator, which consists of a mass m and a rigidity $m\omega_o^2$ (where ω_o is the oscillator's frequency). The oscillator is coupled to a quantum sensor which continuously monitors its coordinate (see chapter VI). The output of the quantum sensor goes into the classical part of the measuring system (not shown in Fig. 8.4), and then on to the observer. The quantum sensor's instantaneous output signal is the sum of the oscillator's instantaneous coordinate $\hat{x}(t)$ and additional noise $\hat{x}_{\text{fluct}}$, which the system superposes on the output. The sensor also exerts a random force $\hat{F}_{\text{fluct}}(t)$ (fluctuational back action) on the oscillator.

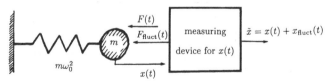

Fig. 8.4 Detection of a classical force by monitoring the coordinate of an oscillator on which it acts.

In this section we shall restrict attention to a sensor that does not use the spectral features of $F(t)$ to optimize its measurements, but instead just monitors the instantaneous value of the probe oscillator's coordinate. (The more general case will be analyzed in the next section.) The spectral density of the internal fluctuations of such a system will be independent of frequency and mutually uncorrelated. It turns out that this is the case, at least within the frequency band of the signal, for all real quantum sensors that are used to monitor probe oscillators.

Since our goal is to compute the ultimate possible sensitivity of such a measuring system, we can assume that (i) the sensor exerts no dynamical (deterministic) back action on the probe oscillator, i.e. the sensor's input susceptibility vanishes; (ii) the monitoring system's noise is the minimum allowed by the uncertainty relations [equation (6.51) is satisfied with an equality rather than inequality], and the influence of all other noise sources are negligible.

We presume, further, that the monitoring system (the sensor and devices that follow it) are permanently coupled to the oscillator. As a result, no matter how small (but nonzero) may be the oscillator's dissipation, during the infinite time that has passed since the coupling began, the oscillator will have completely forgotten its initial state. Reexpressed in the language of the Heisenberg picture, where the state is fixed and the operators evolve, the initial value of the oscillator's position $\hat{x}(-\infty)$ will have been forgotten, and $\hat{x}(t)$ will have been brought into the form

$$\hat{x}(t) = \frac{1}{m\,\omega_o} \int_{-\infty}^{t} \sin\omega_o(t-t')\,\hat{F}_{\text{fluct}}(t')dt' \,, \tag{8.14}$$

where $\hat{F}_{\text{fluct}}(t)$ is the fluctuational back action of the sensor on the probe oscillator. Correspondingly, the sensor's output signal will have the form

$$\tilde{x}(t) = \hat{x}_{\text{fluct}}(t) + \frac{1}{m\,\omega_o} \int_{-\infty}^{t} \sin\omega_o(t-t')\hat{F}_{\text{fluct}}(t')dt'$$

$$+ \frac{1}{m\,\omega_o} \int_{-\infty}^{t} \sin\omega_o(t-t')F(t')dt' \,. \tag{8.15}$$

Here the first term is the noise added by the sensor, the second is the oscillator's position in the absence of the external, classical force, and the third is the effect of that force. The third term can be regarded as the signal, and the first two as the noise from which it must be extracted.

Because the classical part of the measuring device is presumed to perform a direct and completely precise measurement of $\tilde{x}(t)$, we can regard $\tilde{x}$ as classical, and not as a quantum operator. This permits us to discuss the extraction of the signal from the noise in the language of the classical theory of optimal filtering. This theory expresses the confidence of detection as a function of the signal-to-noise ratio,

$$\frac{s}{n} = \int_{-\infty}^{+\infty} \frac{|S_{\text{sig}}(\omega)|^2}{S(\omega)} \frac{d\omega}{2\pi} \,, \tag{8.16}$$

where $S_{\text{sig}}(\omega)$ is the Fourier transform of the signal, and $S(\omega)$ is the spectral density of the noise. In our case, where the signal is the third term of (8.15),

$$S_{\text{sig}}(\omega) = \frac{f(\omega)}{m(\omega_o^2 - \omega^2)} ,\qquad (8.17)$$

aside from an unimportant phase. Here

$$f(\omega) = \int\limits_{-\infty}^{+\infty} F(t)e^{-i\omega t}\,dt \qquad (8.18)$$

is the Fourier transform of the force $F(t)$. Similarly, the spectral density of the noise [first two terms in (8.15)] is

$$S(\omega) = \frac{S_{\text{sensor}}(\omega)}{m^2(\omega_o^2 - \omega^2)^2} ,\qquad (8.19)$$

where

$$S_{\text{sensor}}(\omega) = m^2(\omega_o^2 - \omega^2)^2 S_x + S_F \qquad (8.20)$$

is the spectral density of the all the sensor's fluctuational noise, referred to the probe's input. The spectral densities S_x of x_{fluct} and S_F of F_{fluct} are connected by the condition

$$S_x S_F = \frac{\hbar^2}{4} \qquad (8.21)$$

[equation (6.51) for our situation of the minimum possible noise].

By inserting relations (8.17)—(8.20) into equation (8.16), we obtain

$$\frac{s}{n} = \int\limits_{-\infty}^{+\infty} \frac{|f(\omega)|^2}{S_{\text{sensor}}(\omega)} \frac{d\omega}{2\pi} . \qquad (8.22)$$

The value of this integral depends significantly on the form of the force's spectrum, $f(\omega)$. It should be evident that the signal-to-noise ratio is maximized if the minimum of $S_{\text{sensor}}(\omega)$ occurs at the same frequency as the maximum of $f(\omega)$. This optimal frequency for F, as one easily can show, is

$$\omega_F = \omega_o . \qquad (8.23)$$

Correspondingly, an upper bound on the signal to noise ratio is

$$\frac{s}{n} \le \frac{|f(\omega_F)|^2}{2\pi} \int\limits_{-\infty}^{+\infty} \frac{d\omega}{S_{\text{sensor}}(\omega)} . \qquad (8.24)$$

Equality is achieved when the force has a sufficiently wide bandwidth that

$f(\omega)$ is essentially constant over the range of frequencies that contribute significantly to the integral. (This is the case for bar detectors of gravitational radiation, in which the probe is a mechanical oscillator.)

Evaluation of the integral (8.24) gives

$$\frac{s}{n} = |f(\omega_F)^2| \cdot \{2m(m^2\omega_o^4 S_x + S_F) \cdot$$

$$\cdot [(S_x^{1/2}(m^2\omega_o^4 S_x + S_F)^{1/2} - m\omega_o^2 S_x]\}^{1/2} . \tag{8.25}$$

One can show (we omit the calculations, which are straightforward but very long) that the maximum value of this signal-to-noise ratio, when ω_F is given by (8.23), is achieved when

$$\frac{m\omega_o^2 S_x}{S_F} \gg 1 , \tag{8.26}$$

and is equal to

$$\frac{s}{n} = \frac{|f(\omega_F)|^2}{\hbar\omega_o m} . \tag{8.27}$$

If we choose a signal-to-noise ratio of unity as the criterion for successful detection of the force, then equation (8.27) gives for the minimum detectable force

$$|f(\omega_F)| \geq \sqrt{\hbar\omega_o m} . \tag{8.28}$$

By virtue of relation (8.18), this criterion for detectability is the same as that derived by more elementary techniques in section 8.1 [equation (8.5)].

8.5*   Optimal detection of a classical force

We shall now generalize the analysis of the last section by allowing the quantum sensor's fluctuations to have spectra that depend on frequency. We shall show that, if those sensor noise spectra are "tuned" to match the properties of the probe oscillator, then by the sensor's linear, stationary monitoring of the probe's coordinate, one can extract information about a classical driving force with a sensitivity higher than the standard quantum limit.

We must emphasize that there will be no contradiction between this result and the statements made at the end of section 8.3. Although the sensor is linear and stationarily monitors the probe's coordinate, its spectral features enable it to remove from the signal all information about the coordinate and retain, in the signal passed on to the observer, only

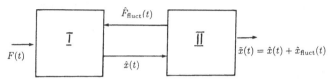

Fig. 8.5 A general scheme for linear detection of a classical force.

information about the driving force.

The scheme of detection that we shall analyze is shown in Fig. 8.5. It is a generalization of the scheme treated in the last section. The harmonic oscillator is replaced by some arbitrary, linear probe object, denoted I in the figure. This object's dynamical behavior is described in the spectral representation by a generalized susceptibility $\chi(\omega)$. The imaginary part of $\chi(\omega)$, according to equation (6.31), characterizes the probe's internal fluctuations. The probe is monitored continuously and stationarily by a linear sensor denoted II in the figure. As in the last section, we shall assume that the sensor's input susceptibility is zero, and its internal fluctuations are the minimum ones permitted by relation (6.51). By contrast with the previous section, we shall not, for now, constrain the noise of the sensor in any other way.

The spectral density of the sensor's total output noise, referred to its input, is easily shown to be

$$S(\omega) = S_{\text{probe}}(\omega) + S_{\text{sensor}}(\omega) , \qquad (8.29)$$

where $S_{\text{probe}}(\omega)$ is the spectral density of the probe's intrinsic fluctuations, and

$$S_{\text{sensor}} = |\chi(\omega)|^2 S_F(\omega) + 2\text{Re}[\chi^*(\omega)S_{xF}(\omega)] + S_x(\omega) \qquad (8.30)$$

is the spectral density of the noise added by the sensor as a result of its three types of internal noise: $S_x(\omega)$, $S_F(\omega)$, and $S_{xF}(\omega)$.

Since the sensor's noise is minimal, equation (6.51) holds with an equality. Inserting into expression (8.30) the value of $S_x(\omega)$ obtained from it, we obtain

$$S_{\text{sensor}}(\omega) = |\chi(\omega)|^2 S_F(\omega) + 2\text{Re}[\chi^*(\omega)S_{xF}(\omega)]$$
$$+ \frac{1}{S_F(\omega)} \{[\text{Re}S_{xF}(\omega)]^2 + [|\text{Im}S_{xF}(\omega)| + \frac{\hbar}{2}]^2\} . \qquad (8.31)$$

We now tailor the form and strength of the noise $\text{Re}S_{xF}(\omega)$ so that, for fixed forms of the other noise, expression (8.31) is minimized. The result is

$$S_{sensor}(\omega) = [\text{Im}\chi(\omega)]^2 S_F(\omega) + 2\text{Im}\chi(\omega)\text{Im}S_{xF}(\omega)$$
$$+ \frac{1}{S_F(\omega)}[|\text{Im}S_{xF}(\omega)| + \hbar/2]^2 . \tag{8.32}$$

The optimal choice of $\text{Re}S_{xF}(\omega)$, which leads to this S_{sensor}, is

$$\text{Re}S_{xF}(\omega) = -S_F(\omega)\text{Re}\chi(\omega) . \tag{8.33}$$

We next tailor the noise $S_F(\omega)$ into the form

$$S_F(\omega) = \frac{1}{|\text{Im}\chi(\omega)|}[|\text{Im}S_{xF}(\omega)| + \hbar/2] , \tag{8.34}$$

which brings the net sensor noise (8.31) to the minimal value

$$S_{sensor}(\omega) = 2|\text{Im}\chi(\omega)|[|\text{Im}S_{xF}(\omega)| + \hbar/2] + 2\text{Im}\chi(\omega)\text{Im}S_{xF}(\omega) . \tag{8.35}$$

Finally, by adjusting $\text{Im}S_{xF}(\omega)$ so that either

$$\text{Im}S_{xF}(\omega) = 0 , \tag{8.36}$$

or

$$\text{Im}\chi(\omega)\cdot\text{Im}S_{xF}(\omega) < 0 , \tag{8.37}$$

we obtain

$$S_{sensor} = \hbar|\text{Im}\chi(\omega)| . \tag{8.38}$$

Notice that this sensor noise is identical to the minimum possible probe noise, $S_{probe}(\omega) = \hbar|\text{Im}\chi(\omega)|$; cf. equation (6.31) as applied to the probe, which has the susceptibility $\chi(\omega)$.

Thus, we have reached the following interesting conclusion: When one is using the system of Fig. 8.5 to measure a classical force, an optimally tailored linear sensor does not have to contribute any more total noise to the measurement than the minimum that can come from the probe. If the probe's internal fluctuations are the smallest permitted by quantum mechanics, then the probe noise and the noise of an optimally tailored sensor are precisely equal. Assuming that this is the case, we now shall compute the minimum detectable force:

The spectral density of the total fluctuations, probe plus sensor, referred to the input of the probe is

$$\frac{S(\omega)}{|\chi(\omega)|^2} = 2\hbar|\omega|H , \tag{8.39}$$

where

$$H = |\omega^{-1}\text{Im}\chi(\omega)| \tag{8.40}$$

is the probe's coefficient of friction. Correspondingly, the signal-to-noise ratio in this case has the form

$$\frac{s}{n} = \int_{-\infty}^{+\infty} \frac{|f(\omega)|^2}{2\hbar |\omega| H} \frac{d\omega}{2\pi} ,$$ (8.41)

where $f(\omega)$ is the Fourier transform of the force that we want to detect. We note that, over the bandwidth of the signal (the band in which $f(\omega)$ is nonnegligible), the coefficient of friction H is typically independent of frequency.

Let us denote by $\overline{\omega}$ and F the mean frequency and amplitude of the classical force, as defined by

$$\overline{\omega} = \frac{\int\limits_{-\infty}^{+\infty} |f(\omega)|^2 d\omega}{\int\limits_{-\infty}^{+\infty} \omega^{-1} |f(\omega)|^2 d\omega} ,$$ (8.42)

$$\xi^2 F^2 \tau_F = \int_{-\infty}^{+\infty} F^2(t) dt .$$ (8.43)

Here τ_F is the duration of the force and ξ is a factor of order unity that depends on its form. Then the minimum detectable force [that which produces a unity signal-to-noise ratio in equation (8.41)] can be expressed as

$$F = \frac{1}{\xi} \sqrt{\frac{2\hbar \overline{\omega} H}{\tau_F}} .$$ (8.44)

This expression differs from the analogous semi-classical formula, which one obtains if one ignores completely the noise introduced by the sensor, only by the factor $\sqrt{2}$.

Thus, we conclude that a linear, stationary, continuously coupled sensor need not impose any significant additional limitations on the sensitivity with which one can monitor an external classical force, beyond those that come from the probe itself—or, rather, it need not do so if one carefully tailors the sensor's noise spectra to match the properties of the probe [while still enforcing the relations (6.51) and (6.50).] As a corollary to this conclusion, we see that this careful tailoring enables the sensor to avoid imposing the standard quantum limit on the measurement accuracy.

8.6* A probe oscillator coupled to a sensor that continuously monitors its number of quanta

As was discussed in section 8.3, one possible method to overcome the standard quantum limit on the measurement of a weak force is by a QND monitoring of the probe's energy. The simple, semiclassical analysis of this method, given in section 8.3, ignored one especially important effect: during a continuous QND monitoring of any observable that has a discrete spectrum of eigenvalues (in this case the energy of the probe oscillator), the response of the probe to an external force is suppressed by the "watchdog" effect (see chapter VII). In section 7.2 it was shown that when the monitoring is exact, then the oscillator completely stops responding to the force. The purpose of this section is to determine what happens when the precision of monitoring is finite, and under what circumstances one can use equation (8.13) for the minimum detectable force.

We begin our analysis with the equation of motion (7.14) for the probe oscillator's density operator. In this equation of motion we must use the probe's Hamiltonian

$$\hat{H}(t) = \hbar\omega_o(\hat{n}+\tfrac{1}{2}) - F(t)\hat{x} \ , \tag{8.45}$$

where $\hat{n}$ and $\hat{x}$ are the oscillator's number operator and coordinate, and $F(t)$ is the classical force. By inserting this Hamiltonian into the equation of motion and identifying the measured operator $\hat{q}$ as the number operator $\hat{n}$, we obtain

$$\frac{\partial\rho_{n'n''}(t)}{\partial t} = -i\,\omega_o(n'-n'')\rho_{n'n''}(t)$$

$$+ \frac{iF(t)}{\sqrt{2m\,\omega\hbar}}\left[\sqrt{n'+1}\rho_{n'+1}(t)_{n''}+\sqrt{n'}\rho_{n'-1\,n''}(t)-\sqrt{n''}\rho_{n'\,n''-1}(t)\right.$$

$$\left.-\sqrt{n''+1}\rho_{n'\,n''+1}(t)\right] - \frac{(n'-n'')^2}{2\tau_o}\rho_{n'n''}(t). \tag{8.46}$$

Here

$$\rho_{n'n''} = \langle n'|\hat{\rho}|n''\rangle \tag{8.47}$$

is the density matrix in the number representation and

$$\tau_o = \hbar^2/\sigma_F^2 \tag{8.48}$$

is the parameter of the sensor, which determines the precision of the monitoring. Using the inequality (7.22), it is easy to show that the variance in the output estimate $\tilde{n}$ of the number of quanta, in the optimal case, is

$$(\Delta\tilde{n})^2 = \frac{\tau_o}{4\tau} \ , \tag{8.49}$$

where τ is the averaging time; cf. equations (7.31) and (7.32).

We shall solve the equation of motion (8.46) by the method of successive approximations, and in doing so shall assume that the force $F(t)$ is small. (This assumption is justified because our goal is to determine the minimum detectable force.) We presume that the oscillator initially is in the energy eigenstate with precisely n quanta:

$$\rho_{n'n''} = \delta_{n'n}\delta_{n''n} \ .$$

Then, in the first approximation,

$$\frac{\partial \rho_{n+1\,n}(t)}{\partial t} + \left[i\,\omega_o + \frac{1}{2\tau_o}\right]\rho_{n+1\,n}(t) = i\sqrt{\frac{n+1}{2\hbar\omega_o m}}F(t) \ ,$$

$$\frac{\partial \rho_{n\,n-1}(t)}{\partial t} + \left[i\,\omega_o + \frac{1}{2\tau_o}\right]\rho_{n\,n-1}(t) = -i\sqrt{\frac{n}{2\hbar\omega_o m}}F(t) \ ,$$

$$\frac{\partial \rho_{n\,n+1}(t)}{\partial t} + \left[-i\,\omega_o + \frac{1}{2\tau_o}\right]\rho_{n\,n+1}(t) = -i\sqrt{\frac{n+1}{2\hbar\omega_o m}}F(t) \ , \quad (8.50)$$

$$\frac{\partial \rho_{n-1\,n}(t)}{\partial t} + \left[-i\,\omega_o + \frac{1}{2\tau_o}\right]\rho_{n-1\,n}(t) = i\sqrt{\frac{n}{2\hbar\omega_o m}}F(t) \ .$$

In the second approximation, the equation for ρ_{nn} takes the form

$$\frac{\partial \rho_{nn}(t)}{\partial t} = \frac{iF(t)}{\sqrt{2\hbar\omega_o m}} \times$$

$$\times\left[\sqrt{n+1}[\rho_{n+1\,n}(t) - \rho_{n\,n+1}(t)] + \sqrt{n}\,[\rho_{n-1\,n}(t) - \rho_{n\,n-1}(t)]\right] \ . \quad (8.51)$$

By inserting into this the time integrals of (8.50) and then integrating over time, we obtain the following probability for the oscillator's state to change:

$$W \equiv 1 - \rho_{nn}(t)\big|_{t\to\infty}$$

$$= \frac{n+\frac{1}{2}}{\hbar\omega_o m}\int_{-\infty}^{+\infty} F(t)F(t')\exp\left[i\,\omega_o\,(t-t') - \frac{|t-t'|}{2\tau_o}\right]dt\,dt' \ . \quad (8.52)$$

This relation takes on the following more transparent form in the spectral representation:

$$W = \frac{n+\frac{1}{2}}{\hbar\omega_o m}\int_{-\infty}^{+\infty} |f(\omega)|^2 G(\omega_o - \omega)\frac{d\omega}{2\pi} \ , \quad (8.53)$$

where $f(\omega)$ is the Fourier transform of the force and

$$G(\omega) = \frac{\tau_o^{-1}}{(2\tau_o)^{-2}+\omega^2} \, . \tag{8.54}$$

From equation (8.53) it follows that the strength of suppression of the oscillator's response is governed by the ratio between τ_o and the effective duration τ_F of the force, where

$$\tau_F^{-1} \equiv \frac{1}{|f(\omega_F)|^2} \int_{-\infty}^{+\infty} |f(\omega)|^2 \frac{d\omega}{2\pi} \, , \tag{8.55}$$

and ω_F is the frequency at which the force's spectrum peaks.

Two limiting cases are of interest:

1. $\tau_o \gg \tau_F$. In this case into (8.53) we must substitute $G(\omega_o -\omega) = 2\pi\delta(\omega-\omega_o)$, and thereby we obtain

$$W = \frac{n+\frac{1}{2}}{\hbar\omega_o m}|f(\omega_o)|^2 \, . \tag{8.56}$$

This formula leads to expression (8.13) for the minimum detectable force.

2. $\tau_o \ll \tau_F$. In this case, and for a force whose frequency is optimized, $\omega_F = \omega_o$, expression (8.13) becomes

$$W = \frac{n+\frac{1}{2}}{\hbar\omega_o m}|f(\omega_o)|^2\frac{4\tau_o}{\tau_F} \, . \tag{8.57}$$

When $\tau_o = 0$ (arbitrarily accurate monitoring), the probe oscillator gets completely frozen into its initial state ($W = 0$), and we have the situation discussed in section 7.1.

Thus, the "watchdog effect" does not affect the sensitivity of the detection system if the parameter τ_o is longer than the duration τ_F of the force. However, the sensitivity is reduced if $\tau_o < \tau_F$. Since, for accurate monitoring, the averaging time τ must be $\leq\tau_o$ [cf. equation (8.48)], this reduction of sensitivity occurs whenever one tries not only to discover whether a force has acted, but also tries to obtain information about the shape of the force.

IX Energetic quantum limitations

9.1 The energy of the probe and the minimum detectable force

There is a remarkable similarity between the expressions for the minimum force that can be detected by a stroboscopic measurement of a probe oscillator's position [equation (8.9)] and by a measurement of its energy [equation (8.13)]. In both cases the sensitivity is inversely proportional to the square root of the fluctuational energy inserted into the probe by the sensor. The same is true for measurements based on a continuous monitoring of one of the oscillator's quadrature amplitudes, as we can see by the following argument:

According to equation (8.4), the minimum detectable force in this case is

$$F \geq \frac{m\omega}{\xi\tau_F}\Delta X_1 \, ,$$

where τ_F is the duration of the force F, ξ is a factor of order unity that depends on the form of the force, m and ω are the mass and eigenfrequency of the probe oscillator, and ΔX_1 is the error in the monitoring of the quadrature amplitude. In view of the uncertainty relation

$$\Delta X_1 \Delta X_2 \geq \frac{\hbar}{2m\omega}$$

[equation (8.3)], this minimum force can be rewritten in the form

$$F \geq \frac{\hbar}{2\xi\tau_F \Delta X_2} \, , \tag{9.1}$$

where ΔX_2 is the perturbation of the other quadrature amplitude due to the measurement process. Since the oscillator's mean energy is

$$\langle E \rangle = \langle E_o \rangle + \langle E_{\text{fluct}} \rangle ,$$

where $\langle E_o \rangle$ is its mean initial energy and

$$\langle E_{\text{fluct}} \rangle = \frac{m\,\omega^2 (\Delta X_2)^2}{2}$$

is the mean energy inserted by the sensor during the measurement, the fluctuational perturbation of X_2 is

$$\Delta X_2 = \sqrt{\frac{2\langle E_{\text{fluct}} \rangle}{m\,\omega^2}} \leq \sqrt{\frac{2\langle E \rangle}{m\,\omega^2}} = \sqrt{\frac{\hbar(n + \frac{1}{2})}{m\,\omega}} .$$

Here n is the mean number of quanta in the oscillator. Inserting this relation into equation (9.1), we obtain, as claimed, that the minimum detectable force is inversely proportional to the square root of $n + \frac{1}{2}$, or equivalently to the square root of the oscillator's energy:

$$F \geq \frac{1}{2\xi\tau_F} \sqrt{\frac{\hbar m\,\omega}{2n + 1}} . \tag{9.2}$$

From this example it is clear that limitations of the form (9.2), (8.9), and (8.14) are caused by the uncertainty principle. To detect the force, it is necessary that the measured probe observable (e.g. X_1) be well defined, and correspondingly, the fluctuations in the observable that is canonically conjugate to the measured one must be large. To enforce this, the sensor must insert a sufficiently large amount of energy into the probe. Thus, the uncertainty principle does not *directly* constrain the sensitivity to the force if the measuring procedure is well designed. Instead, the uncertainty principle couples the ultimate sensitivity to the amount of energy that the measurement must insert into the probe. Since there is always a practical upper limit on the amount of energy the probe can handle, this gives rise to an "energetic limitation" on the sensitivity of the measuring device to any classical force. Moreover, because the practical upper limit is expressible in terms of fundamental physical constants such as the charge of the electron and the masses of the electron and the proton, we can regard the values of these physical constants (which do not enter directly into formal quantum theory) as constraining the precision of quantum measurements in the real world.

By a straightforward computation of the maximum energy the probe can handle, followed by insertion of this result into the expression for the minimum detectable force, one can show that this real-world limitation on the sensitivity is not a severe one. For example, in the case of a probe oscillator, the maximum allowed energy is that whose associated pressure will deform the oscillator so strongly as to exceed its elastic limit, or break it. For the lowest mode of longitudinal oscillation of a solid bar this limit is

$$E < YV(\delta x / x)^2 \, ,$$

where V is the bar's volume, Y is its Young's modulus, and $\delta x / x$ is the maximum strain the bar can stand. For $Y = 10^{12}$ dyn/cm^2, $\delta X / x = 10^{-3}$, $V = 1$ cm^3, and $\omega = 10^6$ sec^{-1}, this maximum energy corresponds to $n < 10^{30}$ quanta of excitation. Correspondingly, the minimum detectable force is 10^{-15} of the standard quantum limit. For electromagnetic resonators, the energetic limits are a little more severe: The maximum energy is limited by electrical breakdown to

$$E < \frac{E_o^2 V}{8\pi} \, ,$$

where V is the effective volume occupied by the electric field and E_o is the breakdown field. For optical microresonators, $V \simeq 10^{-10}$ cm^3, $\omega \simeq 10^{15}$ sec^{-1}, $E_o = 10^4$ statvolts/cm, and consequently the maximum allowed energy corresponds to $n < 4 \times 10^8$, which implies an energetic limit on the sensitivity just four orders of magnitude better than the standard quantum limit.

In examining the practical consequences of these limitations, one must keep in mind that it is not merely necessary for the fluctuational energy to reside in the probe; in fact, the energy must be be inserted into the probe by the sensor during the finite measurement time.

Let us discuss as an example the measurement of the position of a free mass by a shift in the phase of a light wave reflected from it. (This is the principle underlying optical interferometric sensors for small displacements, e.g. the Michelson interferometer sketched schematically in Fig. 9.1.)

The uncertainty relation requires that when the displacement of the mass in Fig. 9.1 is measured with an accuracy Δx, its momentum must be perturbed by an amount no smaller than $\hbar / 2\Delta x$. The source of the perturbation, in this case, is shot-noise fluctuations of the light power. It is easy to show that this shot noise transfers to the mass the following amount of random momentum during the measurement time τ:

$$\Delta P_{\text{perturb}} = \frac{1}{c}\sqrt{2\hbar\omega W \tau} \, ,$$

where ω is the light frequency and W is the light power. From the

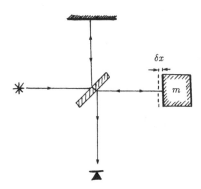

Fig. 9.1 A Michelson interferometer for measuring the displacement δx.

condition

$$\Delta P_{\text{perturb}} \geq \frac{\hbar}{2\Delta x}$$

follows the following minimum required power:

$$W \geq \frac{\hbar c^2}{8\omega(\Delta x)^2\tau} \ .$$

If the desired precision is at the level of the standard quantum limit,

$$\Delta x = \sqrt{\frac{\hbar\tau}{2m}} \ ,$$

then the minimum power is

$$W \geq \frac{mc^2}{4\omega\tau^2} \ .$$

As an example, for typical parameters corresponding to laser-interferometer gravitational-wave detectors, $m = 10^4$ g, $\omega = 3\times10^{15}$ sec^{-1}, $\tau = 10^{-3}$ sec, this power is

$$W \geq 0.8\times10^8 \ \text{watts} \ .$$

The problem of providing this high power is one of the most important in gravitational-wave experiments. To ease the solution, the experimenters use multireflection interferometers with high quality mirrors (mirrors with reflectivities that differ from unity by $1-R \simeq 10^{-4}$—10^{-5}). In this case the high power on the mirror is achieved by accumulating power from the laser inside the interferometer. But even when the number of reflections [approximately equal to $(1-R)^{-1}$] is $\simeq 10^{-5}$, the necessary laser output power is of the order of one.[40]

9.2* Energetic limits on sensitivity: general analysis

In this section we shall generalize equations (8.9), (8.14), and (9.2) to probes with arbitrary structures. In other words, we shall derive a general relationship between the ultimate sensitivity of a quantum probe to an external force and the properties of the probe's quantum state.

It is known that, in the absence of any *a priori* information about the form of the external force, the optimal strategy is to measure a certain operator $\hat{Y}$ of the probe, which minimizes the signal-to-noise ratio

$$\frac{s}{n} = \lim_{F \to 0} \frac{\{\text{Tr}[\hat{Y}(\hat{\rho}_F - \hat{\rho}_o)]\}^2}{\text{Tr}[\hat{\rho}_F \hat{Y}^2] - [\text{Tr}(\hat{\rho}_F \hat{Y})]^2} . \tag{9.3}$$

Here

$$\hat{\rho}_F = \hat{U}_F(t)\hat{\rho}\hat{U}_F^\dagger(t)|_{t \to \infty} \tag{9.4}$$

is the probe's density operator after the external force F has acted on it, $\hat{\rho}$ is the probe's initial density operator, $\hat{U}_F(t)$ is the evolution operator in the presence of the external force, and

$$\hat{\rho}_o = \hat{U}_o(t)\hat{\rho}\hat{U}_o^\dagger(t)|_{t \to \infty} , \tag{9.5}$$

where $\hat{U}_o(t)$ is the operator for free evolution (in the absence of the force F). The operator $\hat{U}_F(t)$ satisfies the Schrödinger equation

$$i\hbar \frac{d\hat{U}_F(t)}{dt} = [\hat{H}_o - F(t)\hat{x}]\hat{U}_F(t) , \tag{9.6}$$

where $\hat{H}_o$ is the probe's Hamiltonian, $F(t)$ is the force that one seeks to detect, and $x(t)$ is that generalized coordinate of the probe on which the force acts. The equation of motion for $\hat{U}_o(t)$ is the same as (9.6), but with $F(t) = 0$.

The signal-to-noise ratio (9.3) is maximal if $\hat{Y}$ satisfies the following equation:

$$\hat{\rho}_o \hat{Y} - \hat{Y}\hat{\rho}_o = 2\frac{\partial \hat{\rho}_F}{\partial F}\bigg|_{F=0} . \tag{9.7}$$

We shall solve this equation in the interaction picture. We begin by writing the evolution operator $\hat{U}_F(t)$ in the form

$$\hat{U}_F(t) = \hat{U}_I(t)\hat{U}_o(t) . \tag{9.8}$$

By inserting this expression in (9.6) we obtain an equation of motion for $\hat{U}_I(t)$:

$$i\hbar \frac{d\hat{U}_I(t)}{dt} = -F(t)\hat{x}(t)\hat{U}_I(t) , \tag{9.9}$$

where

$$\hat{x}(t) = \hat{U}_o^\dagger \hat{x} \hat{U}_o(t)$$

is the coordinate $\hat{x}$ in the interaction picture. Presuming (as is generally the case) that $\hat{x}(t)$ self-commutes twice, i.e. $[[\hat{x}(t),[\hat{x}(t'),\hat{x}(t'')]] = 0$, the solution of (9.9) has the form

$$\hat{U}_I(t) = \exp\left[\frac{i}{\hbar}\int_{-\infty}^t F(t)\hat{x}(t)dt\right] . \tag{9.10}$$

By inserting equations (9.4), (9.5), and (9.8) into (9.7), we obtain

$$\rho\hat{U}_o^\dagger\hat{Y}\hat{U}_o + \hat{U}_o^\dagger\hat{Y}\hat{U}_o\rho = 2\frac{\partial}{\partial F}(\hat{U}_I\rho\hat{U}_I^\dagger) . \tag{9.11}$$

In this relation, the operators $\hat{U}_o$ and $\hat{U}_I$ are supposed to be evaluated for $t\to\infty$; thus, in particular,

$$\hat{U}_I = e^{i\hat{V}} , \tag{9.12}$$

where

$$\hat{V} = \frac{1}{\hbar}\int_{-\infty}^{+\infty}\hat{x}(t)F(t)dt . \tag{9.13}$$

The solution of equation (9.11) can be written in the form

$$\hat{Y} = \frac{2}{i}\hat{U}_o\sum_{\rho\rho'}|\rho>\frac{\rho-\rho'}{\rho+\rho'}<\rho|\hat{V}|\rho'><\rho'|\hat{U}_o^\dagger , \tag{9.14}$$

where ρ and $|\rho>$ are the eigenvalues and eigenstates of the operator $\hat{\rho}$.

By substituting the optimal operator $\hat{Y}$ of (9.14) into (9.3), we obtain the following expression for the optimal signal-to-noise ratio:

$$\frac{s}{n} = 2\sum_{\rho\rho'}\frac{(\rho-\rho')^2}{\rho+\rho'}|<\rho|\hat{V}|\rho'>|^2 . \tag{9.15}$$

To determine the ultimate sensitivity, we now specialize to the case where all fluctuations except pure quantum ones have been excluded from the probe's initial state; i.e. the initial state is pure

$$\hat{\rho} = |\psi><\psi| .$$

In this case (9.15) becomes

$$\frac{s}{n} = <\hat{V}^2>-<\hat{V}>^2 = \frac{1}{\hbar^2}\int_{-\infty}^{+\infty}F(t)\cdot F(t')B(t,t')dtdt' , \tag{9.16}$$

where $B(t,t')$ is the autocorrelation function of the fluctuations in $\hat{x}$:

$$B(t,t') = <\psi|\hat{x}(t)o\hat{x}(t')|\psi> - <\psi|\hat{x}(t)|\psi><\psi|\hat{x}(t')|\psi> . \qquad (9.17)$$

Equations (9.16), (9.17) are the generalization of equations (8.9), (8.14), and (9.2) that we have been seeking. This generalization is valid for a quantum probe of any type whatsoever. This generalization shows that the higher is the sensitivity, the larger must be the fluctuations of x. Physically this comes about because of the uncertainty principle: In order to measure F with high precision, one must have a small uncertainty in the monitored operator $\hat{Y}$ before the measurement begins, and correspondingly one must have a large uncertainty in x.

Suppose, as a particular case, that the probe is an oscillator. Then

$$\hat{x}(t) = \hat{X}_1\cos\omega t + \hat{X}_2\sin\omega t , \qquad (9.18)$$

and

$$\hat{V} = \hat{X}_1\mathrm{Re}f(\omega) - \hat{X}_2\mathrm{Im}f(\omega) , \qquad (9.19)$$

where

$$f(\omega) = \int\limits_{-\infty}^{+\infty} F(t)e^{i\omega t} dt$$

is the Fourier transform of the force at the frequency ω. Inserting expressions (9.18)–(9.20) into (9.16), we obtain

$$\frac{s}{n} = \frac{1}{\hbar^2}<[\Delta\hat{X}_1\mathrm{Re}f(\omega)-\Delta\hat{X}_2\mathrm{Im}f(\omega)]^2> \leq \frac{2n_{\text{fluct}}+1}{\hbar\omega m}|f(\omega)|^2 , \qquad (9.20)$$

where

$$\Delta\hat{X}_{1,2} = \hat{X}_{1,2} - <\hat{X}_{1,2}> ,$$

and

$$\hbar\omega(n_{\text{fluct}}+\tfrac{1}{2}) = \frac{m\omega^2}{2}[<(\Delta\hat{X}_1)^2> + <(\Delta\hat{X}_2)^2>] \qquad (9.21)$$

is the fluctuational energy. Because

$$|f(\omega)| = \xi F \tau_F$$

(where F is the amplitude of the force, τ_F is its duration, and ξ is a factor of order unity that depends on its form, cf. chapter VIII), the force will be detected at the level $s/n = 1$ if

$$F \geq \frac{1}{\xi\tau_F}\sqrt{\frac{\hbar\omega m}{2n_{\text{fluct}}+1}} . \qquad (9.22)$$

Measuring systems based on QND measurements generally have $n_{\text{fluct}} \simeq n$, so equation (9.22) coincides, up to a factor of order unity, with (9.2). The difference between n and n_{fluct} is substantial for states that are nearly classical, i.e. for those with the expectation value of the coordinate far larger than its standard deviation. In particular, for a coherent state $n_{\text{fluct}} = 0$ independently of n, and the best achievable sensitivity is

$$F \geq \frac{1}{\xi \tau_F} \sqrt{\hbar \omega m} \ . \tag{9.23}$$

In other words, the sensitivity is constrained by the standard quantum limit.

9.3* Distinguishing evolutionary paths of a quantum object from each other

The detection of a classical force that acts on a quantum probe is a special case of a more general experimental challenge: To distinguish different evolutionary paths of a quantum object. By distinguishing two such paths one can, for example, detect the interaction of the object with some other quantum system. This more general type of measurement can be analyzed in a manner similar to that of the last section:

Let us limit ourselves to the case of just two paths, which are to be distinguished from each other. We shall denote by H_0 the Hamiltonian that generates the first path, by H_1 the Hamiltonian for the second, and by $\hat{U}_0$, $\hat{U}_1$ the corresponding evolution operators; and we shall assume, for simplicity, that the quantum object begins in a pure initial state ψ_o. The evolution can transform the state into either $\hat{U}_0| \psi_o >$, or $\hat{U}_1| \psi_o >$, depending on which evolution actually occurs. The overlap of these two evolved states is described by the scalar product

$$\gamma = <\psi_o | \hat{U}_o^\dagger \hat{U}_1 | \psi_o > \ . \tag{9.24}$$

The mean probable error in distinguishing the two states from each other is

$$W = \xi_0 \Theta_0 + \xi_1(1 - \Theta_1) \ ,$$

where ξ_0, ξ_1 are the *a priori* probabilities for the two states, and Θ_0, Θ_1 are the probabilities of false alarm and detection. This probability of error is related to the quantity γ as follows:[24]

$$W \geq \frac{1}{2}[1 - (1 - 4|\gamma|^2 \xi_0 \xi_1)^{1/2}] \ . \tag{9.25}$$

Equality is achieved in this relation when the measurements are performed optimally and precisely.

The operator $\hat{R} = \hat{U}_0^\dagger \hat{U}_1$ is analogous to the scattering operator that is used when analyzing a quantum system's evolution in the interaction picture. It satisfies the equation

$$i\hbar \frac{d\hat{R}(t)}{dt} = \hat{H}_I(t)\hat{R}(t) , \qquad (9.26)$$

where

$$\hat{H}_I(t) = \hat{U}_0^\dagger (\hat{H}_1 - \hat{H}_0) \hat{U}_0(t) . \qquad (9.27)$$

Expressing the scattering operator in the form

$$\hat{R} = e^{i\hat{\phi}} ,$$

we obtain

$$|\gamma|^2 = |\int_{-\infty}^{+\infty} e^{i\phi} d\Phi(\phi)|^2 , \qquad (9.28)$$

where $\Phi(\phi)$ is the probability distribution for the eigenvalues ϕ of the operator $\hat{\phi}$ when the object is in its initial state $|\psi_o\rangle$.

It is evident that there are many initial states ψ_o that produce final states orthogonal to each other, i.e. that make $\gamma = 0$ and correspondingly $W = 0$. However, to different initial states correspond different probabilities $\Phi(\phi)$ and correspondingly different moments of the variable ϕ. Below it is shown that the second moment $\Delta\phi$ is connected to the interaction energy associated with the two different evolutions. Thus, it is natural to ask: what is the minimum possible value of $\Delta\phi$ for a given value of $|\gamma|$?

We shall assume that the operator $\hat{\phi}$ has a continuous spectrum of eigenvalues. (The generalization to a discrete or mixed spectrum is not difficult.) In this case we can introduce the probability density $w(\phi) = d\Phi/d\phi$, with normalization

$$\int_{-\infty}^{+\infty} w(\phi) d\phi = 1 , \qquad (9.29)$$

and can write

$$\int_{-\infty}^{+\infty} (\delta\phi)^2 w(\phi) d\phi = (\Delta\phi)^2 , \qquad (9.30)$$

where $\delta\phi = \phi - \langle\phi\rangle$ and $\langle\phi\rangle = \int_{-\infty}^{+\infty} w(\phi)\phi d\phi$. Then equation (9.28) takes the form

$$|\gamma|^2 = |\int_{-\infty}^{+\infty} w(\phi) e^{i\phi} d\phi|^2 \equiv |\int_{-\infty}^{+\infty} w(\phi) e^{i\delta\phi} d\phi|^2 .$$

The value of $|\gamma|$ can be decreased by symmetrizing the probability density $w(\phi)$ with respect to the point $<\phi>$, and such symmetrization does not destroy the normalization (9.29) or change the second moment (9.30). In the special case $|\gamma| = 0$, a distribution of the type

$$w(\phi) = \frac{1}{2}\left[\delta\left[\phi - <\phi> - \frac{\pi}{2}\right] + \delta\left[\phi - <\phi> + \frac{\pi}{2}\right]\right] \qquad (9.31)$$

produces $\Delta\phi = \pi/2$. We can see as follows that when $|\gamma| = 0$, it is not possible to make $\Delta\phi$ smaller than this. Among all distributions normalized by (9.29), the one that produces the absolute minimum $\Delta\phi$ for a given $|\gamma|^2$ can be identified by its extremizing the expression

$$\Lambda = \int_{-\infty}^{+\infty} [\lambda(\delta\phi)^2 + \cos\delta\phi]w(\phi)d\phi ,$$

where λ is a Lagrange multiplier. The fact that $\Lambda \geq \lambda\phi_o^2 + \cos\phi_o$ for all normalized choices of $w(\phi)$ (where ϕ_o is the location of the minimum of $\lambda x^2 + \cos x$) and the fact that equality is achieved when

$$w(\phi) = \frac{1}{2}[\delta(\phi - <\phi> - \phi_o) + \delta(\phi - <\phi> + \phi_o)] \qquad (9.32)$$

imply that expression (9.32) must be the optimal distribution and that $\Delta\phi = \phi_o$ and $\gamma = \cos(\Delta\phi)$. If $\Delta\phi < \pi/2$, then $\gamma > 0$; if $\Delta\phi = \pi/2$, then $\gamma = 0$ and the distribution (9.32) is the same as (9.31). Thus, a necessary condition for error-free distinguishing between the two evolutionary paths is

$$\Delta\phi \geq \pi/2 . \qquad (9.33)$$

If this condition is not satisfied, then

$$|\gamma| = \cos\Delta\phi . \qquad (9.34)$$

Notice that the probability of error (9.25), being a function of $|\gamma|$, is determined only by the variance of ϕ and is unaffected by its mean value.

The operator $\hat{\phi}$ is uniquely related to the operator $\hat{H}_I$. If the self-commutator of this $\hat{H}_I$ is a c-number, i.e. if

$$[[\hat{H}_I(t), \hat{H}_I(t')], \hat{Q}] = 0 , \qquad (9.35)$$

then

$$\hat{\phi} = -\frac{1}{\hbar}\int_0^\tau \hat{H}_I(t')dt' , \qquad (9.36)$$

where τ is the time of evolution. By taking account of this relation, one can reexpress the inequality (9.33) as

$$\Delta \overline{H}_I \cdot \tau \geq \frac{\pi}{2} \hbar \, , \qquad (9.37)$$

where $\Delta \overline{H}_I$ is the uncertainty in the mean time difference between the system's energies along its two different evolutionary paths. In situations where one is trying to detect an interaction between two quantum objects, this $\hat{H}_I$ will be the interaction energy.

X Devices for measuring small mechanical displacements

10.1 Parametric transducer for mechanical displacements

Most modern instruments used to measure the absolute distance x or changes in distance δx between macroscopic bodies are one variant or another of parametric transducers. A key element in any parametric transducer is an optical or radio frequency electromagnetic auto-oscillator (self-excited oscillator), with high frequency stability and a low level of power fluctuations. The transducer is constructed so that the x or δx to be measured produces a change in the frequency of the auto-oscillator itself or the frequency of a high-Q resonator that the auto-oscillator drives, or it produces a change of the optical length of an electromagnetic beam generated by the auto-oscillator (e.g. in laser or radio ranging systems or in interferometers), which length change is measured via a resulting change in the beam's power or phase.

To obtain high sensitivity from such a parametric transducer one typically must achieve *i)* high frequency stability of the auto-oscillator, and *ii)* a flow of a large amount of energy $E = W \cdot \tau$ "through" the transducer during a measuring time τ.

The first of these conditions leads to an accuracy

$$\frac{\Delta x}{x} \geq \left[\frac{\Delta \omega}{\omega} \right]_{a-o} . \tag{10.1}$$

Here Δx is the error in the measurement of x or δx, and in the case of an absolute measurement of x, $(\Delta \omega / \omega)_{a-o}$ is the fractional instability of the auto-oscillator's frequency, while in the case of measurements of changes δx of x, $(\Delta\omega/\omega)_{a-o}$ is the oscillator's short-term instability—i.e. the fractional fluctuations of ω in those parts of the spectrum over which one wishes to measure δx. In the latter case, the accuracy can be improved by using special compensation techniques in the transducer.

Condition *ii)* (high energy flow through the transducer) leads, after some simplification, to an accuracy

$$\frac{\Delta x}{x} \gtrsim \frac{1}{\kappa}\sqrt{\frac{k_B T_N}{E}} \, , \qquad (10.2a)$$

where T_N is the amplifier's noise temperature; or, when quantum effects are important (when the noise temperature is close to $\hbar\omega_e/2k_B$), to

$$\frac{\Delta x}{x} \gtrsim \frac{1}{\kappa}\sqrt{\frac{\hbar\omega_e}{2E}} \, . \qquad (10.2b)$$

Here ω_e is the electromagnetic frequency, and κ is a dimensionless factor that depends on the type of transducer. For a capacity transducer

$$\kappa = Q_e \, ,$$

where Q_e is the quality factor of the transducer's resonator; for a ranging device such as those used to track interplanetary spacecraft,

$$\kappa = \omega_e \tau_x \, ,$$

where $\tau_x = x/c$; for a device based on a Fabry-Perot resonator,

$$\kappa = \frac{\omega_e \tau_x}{1-R} = Q_{\text{optical}} \, ,$$

where R is the reflectivity of the resonator's mirrors, and Q_{optical} is its quality factor; and for a multipass Michelson interferometer,

$$\kappa = \omega_e \tau_x N \, ,$$

where N is the number of passes of the light between the mirrors. If the frequency ω_e is in the optical region, then always $\hbar\omega_e \gg k_B T$, so in making accuracy estimates one must use equation (10.2b)—which, however, is valid only if there is no substantial excess noise in either the auto-oscillator or the receiver.

Condition (10.2b) corresponds to the case when the electromagnetic radiation is used in its "natural," coherent state, i.e. a state in which the photons are distributed independently of each other and therefore give Poisson-distributed shot noise. By preparing the quantum state of the electromagnetic wave in special ways (see chapter XII), one can reduce substantially the amount of energy E required for a given accuracy.

Table 1. Precisions achieved in various types of parametric transducers

	x, cm	δx, cm	$\dfrac{\delta x}{x}$; $\dfrac{\delta\varepsilon}{\varepsilon}$	Type of Measurement	Ref
1	3×10^{13}	2×10^2	6×10^{-12}	absolute measurements of x; radio ranging to spacecraft with active transponder	[41]
2	3.8×10^{10}	1.7	5×10^{-11}	absolute measurements of x; laser ranging to the moon	[42]
3	4×10^3	1.2×10^{-14}	3×10^{-18}	differential measurements of x; $\Delta f \simeq 10^3$ Hz, $\bar{f} \simeq 10^3$ Hz; optical Fabry-Perot resonator, Michelson interferometer	[43] [44]
4	3×10^{-4}	6×10^{-17}	2×10^{-13}	differential measurements of x; $\Delta f =1$ Hz, $\bar{f} \simeq 8$ kHz; capacitive cryogenic transducer	[45]
5	3×10^{-5}	3×10^{-15}	1×10^{-10}	differential measurements of ε; $\Delta f \simeq 0.1$ Hz, $\bar{f} =0.1$ Hz; capacitive cryogenic transducer	[46]

Table 1 shows five examples of the accuracy achieved in precise measurements of absolute distance x, of small distance changes δx, and of small changes in the dielectric susceptibility ε of the medium through which an electromagnetic wave propagates. (It should be evident that a change of ε has the same effect on an electromagnetic wave as a change of x.) The first two examples in Table 1 are absolute measurements of x; the last three are differential measurements of small $\delta x/x$ and $\delta\varepsilon/\varepsilon$.

The significance of these achieved resolutions and precisions can be summarized by the statement that, despite the very small Δx, $\delta x/x$, and $\delta\varepsilon/\varepsilon$ that have been obtained, there remains substantial room for further improvements in both the absolute and the differential measurements. The experiments thus far have not fully used the substantial existing reserve of frequency stability for existing auto-oscillators; and in some cases they did not use the largest values of the factor κ that are available today for the chosen types of parametric transducers. In many cases, the accuracy was determined not by the conditions (10.1) or (10.2), but instead by other effects that might be partially or wholly removed in the future. One can expect that in the coming years new experiments will achieve values of Δx and $\delta x/x$ that are several orders of magnitude smaller than those in Table 1.

It is worth noting that there is a level of frequency stability beyond which further attempts at improvements will be confronted by purely quantum (measurement) effects, analogous to those discussed in section 1.4. In particular, one can show (see details in Reference 47) that there exists a standard quantum limit for the frequency stability of an auto-oscillator, $(\Delta\omega/\omega)_{SQL}$, analogous to the standard quantum limits for the coordinate of a free mass or oscillator [equations (1.23) and (1.24)]. If the auto-oscillator's resonator has a volume V and this volume is filled with a rigid dielectric with Young's modulus Y, then $(\Delta\omega/\omega)_{SQL}$ can be written in the simple form

$$\left[\frac{\Delta\omega}{\omega}\right]_{SQL} \simeq \sqrt{\frac{\hbar}{VY\tau}}\,, \qquad (10.3a)$$

where τ is the averaging time. This limit can be achieved only if the power in the auto-oscillator is optimal

$$W_{optimal} \simeq \frac{VY\omega_e}{Q_e^2}\,. \qquad (10.3b)$$

Equations (10.3a,b) include as a special case the well-known Schawlow-Townes formula.[48]

In concluding this section, let us emphasize two important things:

a)   The smallest value of Δx that has been achieved (see, e.g., item 4 in Table 1) is close to the standard quantum limit of section 1.5. This is one basis for optimistic prognoses that have been made for programs aimed at "beating" this limit.

b)   Condition (10.2b), as was mentioned above, is valid only if the radiation generated by the auto-oscillator is in a coherent quantum state; and this condition can be relaxed by the use of other quantum states (see below). Thus, there exists another reserve for increases of sensitivity, but this reserve can be realized only after methods of preparing and detecting these non-coherent quantum states have been developed.

10.2 Capacity transducer

The remainder of this chapter is devoted to a detailed discussion of a typical example of a parametric transducer for measuring differential displacements. This example, the capacity transducer, is used in measurements of very high precision; cf. items 4 and 5 in Table 1. We have chosen this example for the following reasons:

First, the capacity transducer is one of the most sensitive sensors of all those presently used to measure small displacements. It has achieved a level of precision for mechanical position measurements of order the standard quantum limit. Second, a capacity transducer can be operated in an interesting variety of regimes, including some that satisfy the criteria for QND measurements. Third, the principle underlying a capacity transducer is extremely simple and transparent, and correspondingly a theoretical analysis of measurements made with it is typically not complicated.

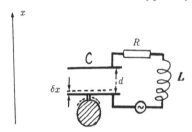

Fig. 10.1 The scheme of a capacity transducer.

The traditional variant of a capacity transducer is an electrical LC circuit, with one of the plates of the capacitor rigidly coupled to the mass whose position x is to be measured. The mass and its attached plate are free to move along the x-axis as shown in Fig. 10.1. A displacement of the mass changes the gap between the capacitor plates and thereby changes the eigenfrequency ω_e of the electrical circuit. Almost always two strong inequalities are satisfied: First, the motion of the mechanical object is far slower than the oscillations in the electrical circuit:

$$1/\tau_M \ll \omega_e \;, \qquad\qquad (10.4a)$$

where τ_M is the characteristic timescale for changes in the position x. (If, for example, the mass is part of a mechanical oscillator, then $\tau_M \simeq \omega_M^{-1}$, where ω_M is its eigenfrequency.) Usually the $1/\tau_M$ on the left side of (10.4a) is 6 to 7 orders of magnitude smaller than the ω_e on the right. Second, the range over which x changes is tiny compared to the mean distance d between the plates:

$$\delta x/d \ll 1 \;. \qquad\qquad (10.4b)$$

An auto-oscillator drives the electrical circuit at a "pump frequency" ω_p which is close to its eigenfrequency ω_e, and the displacement-induced changes of ω_e cause changes in the amplitude and phase of the circuit's pump-induced oscillations. By attaching an amplitude or phase detector, one can observe these changes, and thereby infer details of the mechanical

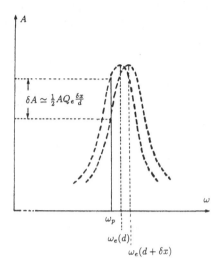

Fig. 10.2 The "tuned-to-the-slope" regime of operation of a capacity transducer.

object's displacement.

It is important to note that, when x changes, the circuit's oscillations monitor it not instantaneously, but rather with a time delay of order the circuit's relaxation time τ_e^*. Thus, to monitor the position x, it is necessary that the object change its motion slowly:

$$\tau_M > \tau_e^* . \tag{10.5}$$

The regime of operation in which this is the case is called the "stationary regime." In the simplest case, the circuit's oscillations are monitored using an amplitude detector. To have a maximal sensitivity of the electrical oscillations' amplitude A to changes of x, one must pump the oscillator at the point of maximum slope of its resonance curve, i.e. one resonance half-width away from the resonance peak. This is the so-called "tuned-to-the-slope" regime of operation; cf. Fig. 10.2. (Obviously, one can operate simultaneously in the stationary regime and the tuned-to-the-slope regime.) In this regime the fractional amplitude change $\delta A / A$ produced by a displacement δx is

$$\frac{\delta A}{A} \simeq \frac{1}{2} Q_e \frac{\delta x}{d} . \tag{10.6a}$$

The advantage of the tuned-to-the-slope regime is simplicity of the transducer. A disadvantage is the fact that the transducer distorts the dynamics of the object. Specifically: There is an electrostatic force between the plates proportional to the squared amplitude A^2 of the electrical voltage. The dc component of this force can easily be compensated, but in the tuned-to-the-slope regime, the voltage amplitude and thus the force depend substantially on x. This dependence is equivalent to adding a certain rigidity to the object (the so-called "dynamical electromagnetic rigidity"). There is a delay τ_e^*, however, in the response of the amplitude A and thence the rigidity to a change in x. The forces produced by such a time-retarded rigidity are known to amplify or de-amplify the oscillations of the mechanical mass (depending on the sign of the rigidity).

These complicating effects do not occur if the pump frequency is tuned to the peak of the resonance curve, i.e. if $\omega_p = \omega_e$, because in this "tuned-to-the-peak" regime, the amplitude of oscillation depends only weakly on x (presuming, of course, that δx is small enough that it does not drive the peak of the resonance curve significantly away from ω_p, i.e. presuming $\delta x / d < 1/\omega_e \tau_e^*$). In the tuned-to-the-peak regime, an amplitude detector is useless; in its place one must use a phase detector. The sensitivity of the oscillator's phase ϕ to changes of x is maximal at the peak of the resonance:

$$\delta\phi \simeq Q_e \frac{\delta x}{d} \; ; \tag{10.6b}$$

cf. Fig. 10.3.

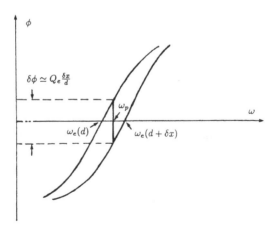

Fig. 10.3 The "tuned-to-the-peak" regime of operation of a capacity transducer.

It follows from the general analysis of chapter VI that the sensitivity of a capacity transducer in the stationary regime is constrained by the standard quantum limit, in principle. However, there exist other, more sophisticated regimes for a capacity transducer, in which the standard quantum limit can be beaten (at the price of much more severe restrictions on the quality factor of the transducer's circuit, the pump power, etc.). Sections 10.5—10.7 are devoted to these regimes. The most interesting of them is the regime of so-called "two-side-band pumping," in which the pumping electric field in the capacitor varies harmonically at the frequency ω_M of the mechanical oscillations of the measured displacement x. This regime is a practical realization of the scheme for measuring the quadrature amplitude X_1, discussed in section 8.3 and Fig. 8.3. The gain in this case compared to the standard quantum limit is of order $\sqrt{\omega_M \tau_e^*}$, so to achieve this gain imposes rather severe conditions on the quality factor of the capacity transducer's electromagnetic resonator. For a detailed analysis, see section 10.7.

10.3 Fluctuations in a capacity transducer in the stationary regime

In chapter VI, using a very general approach, we derived constraints on the noise of any linear device for monitoring the position of a quantum object. Since the capacity transducer in the stationary regime is a linear device, its noise must satisfy those constraints. On the other hand, because of the capacity meter's simplicity, one can compute directly its noise characteristics. Such a computation will make clear the connection between the generalized sources of noise x_{fluct} and F_{fluct} of chapter VI, and the real physical sources of fluctuations in a concrete measuring device.

In the capacity transducer sketched schematically in Fig. 10.1, the only source of fluctuations is the resistance R. The spectral density of the random voltage V_R produced by this resistance, according to the fluctuation-dissipation theorem, is

$$S_R = 2k_B T_e R \ . \tag{10.7}$$

Here k_B is Boltzmann's constant, and T_e is the resistance's effective noise temperature—which is equal to the physical temperature T if $k_B T \gg \hbar\omega_e/2$, and is equal to $\hbar\omega_e/2k_B$ in the opposite case, $k_B T \ll \hbar\omega_e/2$. The exact formula for T_e is

$$T_e = \frac{\hbar\omega_e}{2k_B}\coth\frac{\hbar\omega_e}{2k_B T} \ .$$

Recall that the spectral density S_R determines the mean square voltage fluctuations in any range $\Delta\omega$ of (angular) frequencies:

$$\langle V^2 \rangle_{\Delta\omega} = S_R \frac{\Delta\omega}{\pi} \ .$$

It is important to note the physical nature of the R in these relations: In order for the transducer to be in the stationary regime, the electrical circuit's relaxation time

$$\tau_e^* = \frac{2}{C\omega_e^2 R}$$

(where C is the capacitance) must be less than τ_M, and correspondingly, the circuit's resistance must be at least as large as

$$R > \frac{2}{C\omega_e^2 \tau_M} .$$

However, to optimize the transducer's sensitivity, one must keep its fluctuations as small as possible. This can be done, e.g., by using some sort of artificial damping to keep τ_e^* small. When this is done, then the resistance R that produces the fluctuations of equation (10.7) is not the source of the oscillator's damping, but instead is the input resistance of the detector that measures the circuit's oscillations, and the voltage fluctuations are that detector's back action on the circuit. (The detector is called a sensor in some earlier chapters.)

Thus, the voltage across the transducer's capacitor plates (when $\delta x = 0$) is the sum of the voltage $V_p(t)$ induced directly by the pump and the fluctuations $V_{\text{fluct}}(t)$ produced by the voltage noise $V_R(t)$. This voltage produces an attractive force between the plates given by

$$F(t) = \frac{C}{2d}[V_p(t) + V_{\text{fluct}}(t)]^2$$

$$= \frac{C}{2d}V_p^2(t) + \frac{C}{d}V_p(t)V_{\text{fluct}}(t) + \frac{C}{2d}V_{\text{fluct}}^2(t) .$$

Since, in general, $V_p \gg V_{\text{fluct}}$, the third term can be neglected. The first term is a regular, known force of attraction that can be taken into account in the data analysis or can be compensated. The second term remains. The mechanical system (measured object) responds only to the low-frequency part of this noise force's spectrum, i.e. to

$$F_{\text{fluct}}(t) = \frac{C}{d}\overline{V_p(t) \cdot V_{\text{fluct}}(t)} ,$$

where the bar denotes an average for a time τ given by $\tau_e^* \ll \tau \leq \tau_M$). This force is the fluctuational back action of the capacity transducer on the measured object. The spectral density of this force is given by (we omit the derivation, which is of no essential interest):

$$S_F = \frac{CE^2 Q_e}{\gamma^2 + 1} \cdot \frac{k_B T_e}{\omega_e} , \tag{10.8}$$

where E is the amplitude of the pumped electric field in the capacitor gap, and

$$\gamma = (\omega_p - \omega_e)\tau_e^*$$

is the detuning of the pump frequency from the circuit's eigenfrequency in units of the circuit's resonance width.

In accord with the results of chapter VI, the transducer's output signal includes not only the useful signal which carries the information about x, but also a noise component. In the case of amplitude detection, the output signal is the power W entering the detector, so

$$W = Kx + W_{\text{fluct}} = K(x + x_{\text{fluct}}) \ .$$

Here K is the transfer coefficient for converting a change of x into a change of the power,

$$K = \frac{\delta W}{\delta x} \ ,$$

and

$$x_{\text{fluct}} = \frac{W_{\text{fluct}}}{K} \tag{10.9}$$

is the noise referred to the transducer's input. In fact, the flux of energy into the detector (which is given by

$$W = \frac{RI^2}{2}$$

where I is the amplitude of the circuit's current) consists of individual quanta with energies close to $\hbar\omega_e$, and correspondingly it is accompanied by shot noise with the spectral density

$$S_W = \hbar\omega_e W \ .$$

This is the absolute minimum noise, valid only when there are no other sources of fluctuations. If there are additional fluctuations (e.g. of thermal origin), then

$$S_W = 2k_B T_e W \ ,$$

where

$$k_B T_e \geq \frac{\hbar\omega_e}{2} \ .$$

Strictly speaking, the effective temperature T_e in these formulae is different from that in equation (10.8), but their values are close to each other, and for estimates of the ultimate sensitivity, they can be regarded as equal. (Only their geometric mean enters into the final formulae.)

The spectral density S_x of the noise x_{fluct} referred to the transducer's input is given, according to equation (10.9), by

$$S_x = \frac{1}{K^2} S_W \ .$$

The value of K can be computed in the following way. When the position x of the capacitor plate changes by an amount $\delta x \ll d$, then the circuit's frequency changes by

$$\delta \omega = \frac{\omega_e \delta_x}{2d}$$

and the amplitude of the circuit's oscillations changes correspondingly by

$$\delta A = \frac{\gamma Q_e A}{\gamma^2 + 1} \frac{\delta x}{d} \ .$$

This causes the power going to the detector to change by

$$\delta W = \frac{C \omega_e E^2 \gamma d}{\gamma^2 + 1} \delta x \ .$$

Therefore,

$$K = \frac{C \omega_e E^2 \gamma d}{\gamma^2 + 1} \ .$$

This value of K implies that, when amplitude detection is used,

$$S_x = \frac{2(\gamma^2 + 1)^2}{CE^2 Q_e \gamma^2} \cdot \frac{k_B T_e}{\omega_e} \ . \tag{10.10}$$

The product of the spectral densities (10.8) and (10.10) is

$$S_x \cdot S_F = \frac{\gamma^2 + 1}{\gamma^2} \cdot \left[\frac{k_B T_e}{\omega_e} \right]^2 \ . \tag{10.11}$$

One can analyze a circuit that is monitored by phase detection rather than amplitude detection in the same way. The spectral density of the phase fluctuations is given by the Schawlow-Townes formula[48]

$$S_\phi = \frac{k_B T_e}{2W} \ .$$

A change of the object's position by the amount δx produces the following change of phase in the circuit's oscillations:

$$\delta \phi = \frac{Q_e}{\gamma^2 + 1} \frac{\delta x}{d} \ .$$

Therefore, in this case

$$S_x = \left[\frac{\delta x}{\delta \phi} \right]^2 S_\phi = \frac{(\gamma^2+1)^2}{CE^2 Q_e} \cdot \frac{k_B T_e}{\omega_e} . \tag{10.12}$$

This implies that, when phase detection is used, the product of the spectral densities S_x and S_F is

$$S_x \cdot S_F = (\gamma^2+1) \left[\frac{k_B T_e}{\omega_e} \right]^2 . \tag{10.13}$$

Let us compare the noise formulae (10.11) and (10.13). For amplitude detection [formula (10.11)], the product of the spectral densities decreases as γ increases, asymptotically approaching $(k_B T_e / \omega_e)^2$ in the limit $\gamma \to \infty$. This result is easy to explain: In general, x influences both the circuit's phase and its amplitude. When amplitude detection is used, some portion of the total information encoded in the oscillations gets lost. Correspondingly, the transfer coefficient is smaller than the maximum possible, and the noise referred to the input is larger than the minimum possible. As γ is increased, the influence of x on the phase is reduced, and in the limit $\gamma \to \infty$, amplitude detection becomes the optimal procedure. Correspondingly, the right-hand side of (10.11) asymptotes to its minimum value $(k_B T_e / \omega_e)^2$.

When phase detection is used, the situation is analogous with only one difference: The phase detection becomes optimal not when $\gamma \to \infty$, but rather when $\gamma=0$, because the amplitude of the oscillations is unaffected by small changes of x when the detuning is zero (when the pumping is "tuned-to-the-top"). Correspondingly, the right-hand side of equation (10.13) is minimized when $\gamma = 0$.

In principle, when the signal is encoded in both the amplitude and the phase (in a combination that depends on the amount of detuning), it is possible to make optimal measurements using a combination scheme involving both amplitude detectors and phase detectors. One can show that in this case

$$S_x = \frac{\gamma^2+1}{CE^2 Q_e} \frac{k_B T_e}{\omega_e} , \tag{10.14}$$

and the product of the spectral densities is

$$S_x S_F = \left[\frac{k_B T_e}{\omega_e} \right]^2 , \tag{10.15}$$

independently of the pumping frequency. In the pure quantum limit, when

$$k_B T_e = \frac{\hbar \omega_e}{2} ,$$

this relation takes the form

$$S_x S_F = \frac{\hbar^2}{4} . \tag{10.16}$$

Thus, using the method of a direct calculation of the spectral densities of the noises of a concrete measuring system, we have obtained the same relation as was derived from the general theory of linear measurements in chapter VI; cf. equation (6.7).

10.4 Capacity transducer used to detect weak forces: stationary regime

The record resolutions for a capacity transducer were achieved in experiments designed to detect gravitational waves. In these experiments the goal was not to detect the position of an object, but rather to register the action of a classical external force on the probe object: a mechanical oscillator. Two fundamental effects limit the sensitivity of such experiments: thermal fluctuations in the object itself, associated with its internal dissipation; and back action of the capacity transducer on the object.

Consider, first, the thermal fluctuations. The spectral density of their random force is given by

$$S_{\text{th}} = \frac{4k_B T_M m}{\tau_M^*} ,$$

where m is the probe oscillator's mass, τ_M^* is its relaxation time, and T_M is its temperature. To detect the external force in the midst of this background, the force's amplitude must be at least as strong as

$$F \geq \frac{1}{\xi} \sqrt{\frac{S_{\text{th}}}{\tau_F}} = \frac{1}{\xi} \sqrt{\frac{4k_B T_M m}{\tau_M^* \tau_F}} . \tag{10.17}$$

Here τ_F is the duration of the force and ξ is a numerical factor of order unity that depends on the force's shape. Equation (10.17) is the "potential sensitivity" of the probe oscillator.

It is evident from this equation that the potential sensitivity can be increased by decreasing the temperature T_M of the probe oscillator, and by decreasing its thermal dissipation (i.e. increasing τ_M^*). When the thermal fluctuations have been reduced sufficiently, the sensitivity begins to be determined by the fluctuating back action of the capacity transducer.

Let us examine the maximum sensitivity that can be achieved, and how one might achieve it, when the probe oscillator has small enough internal friction and low enough temperature that we can ignore thermal fluctuations, and when the capacity transducer operates in the stationary regime. The condition of stationarity in this case can be written as

$$\tau_e^* \omega_M \ll 1 \, ,$$

where ω_M is the eigenfrequency of the probe oscillator. If this condition is satisfied, the transducer can monitor the probe oscillator's position x. According to the general analysis of chapter VIII, the sensitivity in this case is constrained by the standard quantum limit (8.5). It is evident that this limit can be achieved only if the transducer's noises are of pure quantum origin, i.e.

$$T_e \leq \frac{\hbar \omega_e}{2 k_B} \, .$$

If there is additional, classical noise, then [as one can see by comparing equations (6.7) and (10.15)] in the formula for the minimum detectable force one must replace $\hbar/2$ by $k_B T_e / \omega_e$, thereby obtaining

$$F_{\min} > \frac{1}{\xi \tau_F} \sqrt{\frac{k_B T_e \omega_M m}{\omega_e}} \, . \tag{10.18}$$

The level of sensitivity (10.18) can be achieved only if the transducer's circuit is being pumped sufficiently strongly. Specifically:

The level of sensitivity (10.18) corresponds [according to equation (8.4)] to an error

$$\Delta x = \sqrt{\frac{k_B T_e}{m \, \omega_M \omega_e}}$$

in monitoring the coordinate of the probe oscillator. This precision can be achieved with an averaging time τ only if the spectral density of the noise x_{fluct} satisfies

$$\sqrt{\frac{S_x}{\tau}} \leq \sqrt{\frac{k_B T_e}{m \, \omega_M \omega_e}} \, .$$

Inserting this into expression (10.14) for S_x, we obtain the following constraint on the strength of the pumping electric field between the capacitor plates:

$$E^2 \geq \frac{m \, \omega_M \, (\gamma^2 + 1)}{C \tau Q_e} \, . \tag{10.19}$$

This condition is a rather challenging one: For the parameters $m = 10^4$ g, $\omega_M = 10^4$ sec^{-1}, $C = 10$ cm, $\tau = 10^{-3}$ sec, $Q_e = 10^5$, and $\gamma = 0$, it demands $E \geq 3 \times 10^4$ volt/cm $= 10^2$ esu.

In section 8.2 it was shown that one can beat the standard quantum limit for an oscillator by measuring it stroboscopically instead of stationarily; i.e., by coupling a transducer to the oscillator's coordinate briefly, at moments of time separated by half the oscillator's period, π/ω_M. The stroboscopic procedure can be realized using a capacity transducer by pumping the transducer's circuit with short impulses of duration

$$\tau_{imp} \ll 2\frac{\pi}{\omega_M} \; .$$

The impulses, however, must be long enough to give the transducer's circuit time enough to become stationary:

$$\tau_{imp} \gg \tau_e^* \; .$$

This permits us to treat the capacity transducer as operating in the stationary regime while its pumping is on.

Since the time interval τ_{imp} is much shorter than the oscillator's period, the oscillator behaves as though it were a free mass. In particular, the minimum possible error in monitoring its coordinate is the free-mass standard quantum limit (1.23). If the noise temperature of the capacity transducer used to monitor the coordinate exceeds the quantum level $\hbar\omega_e/2k_B$, then the minimum error is the correspondingly modified free-mass standard quantum limit

$$\Delta x = \sqrt{\frac{k_B T_e \tau_{imp}}{m\,\omega_e}} \; .$$

Because the oscillator's wave function spreads immediately after each measurement of this accuracy, but then reconstructs its sharp concentration at the time of the next measurement, this is the relevant accuracy for one stroboscopic measurement after another. Inserting this expression into equation (8.7), we obtain the minimum detectable force

$$F_{min} = \frac{1}{\xi\tau_F}\sqrt{\frac{k_B T_e m \tau_{imp}}{\omega_e}} = F_{SQL}\cdot\sqrt{\frac{2k_B T_e}{\hbar\omega_e}}\cdot\sqrt{\omega_M \tau_{imp}} \; , \qquad (10.20)$$

where F_{SQL} is the standard quantum limit (8.5) for the oscillator. The sensitivity, therefore, is higher the smaller is the measurement time τ_{imp}. For sufficiently small τ_{imp}, it can beat the oscillator's quantum limit, even when T_e is somewhat larger than $\hbar\omega_e/2k_B$. "Sufficiently small," according to equation (10.20), means

$$\omega_M \tau_{imp} < \frac{\hbar \omega_e}{2 k_B T_e} .$$

The main technical problem in realizing such a scheme is to excite the transducer's circuit to a high level during the short time τ_{imp}. How high?: The spectral density S_x is limited by

$$\sqrt{\frac{S_x}{\tau_{imp}}} \lesssim \sqrt{\frac{k_B T_e \tau_{imp}}{m \omega_e}} .$$

From this and equation (10.14) follows the following expression for the electric field E that must be stroboscopically pumped into the transducer's capacitor:

$$E^2 \geq \frac{m(\gamma^2+1)}{C \tau_{imp}^2 Q_e} . \tag{10.21}$$

For example, if $m = 10^4$ g, $C = 10$ cm, $\tau_{imp} = 10^{-4}$ sec, $Q_e = 10^5$, $k_B T_e \simeq \hbar \omega_e / 2$, $\gamma = 0$, then the capacitor's impulsively excited field must be $E \geq 10^3$ esu $= 3 \times 10^5$ volt/cm.

10.5* Capacity transducer: nonstationary regime

This section and the next two are devoted to a capacity transducer operating in a nonstationary regime—i.e., in a regime where the oscillations in the transducer's circuit depend not on the instantaneous value of the measured coordinate x, but on the evolution of x during the relaxation time τ_e^*. The analysis of this regime is more complicated than that of the stationary regime. This section will discuss several conditions that permit the analysis to be simplified. Then the following two sections will analyze two specific variants of nonstationary measurements.

The equation of motion for the transducer's circuit has the form

$$L \frac{d^2 q_\Sigma}{dt^2} + R \frac{dq_\Sigma}{dt} + \frac{q_\Sigma}{C} \left[1 - \frac{x(t)}{d} \right] = V_p(t) + V_{fluct}(t) , \tag{10.22}$$

where L, C, and R are the circuit's inductance, capacity, and resistance, q_Σ is the charge on the capacitor, $V_p(t)$ is the pumping voltage, and $V_{fluct}(t)$ is a fluctuating e.m.f. that consists of the thermal noise and the back-action fluctuational influence of the sensor which monitors the transducer's circuit.

The circuit's oscillations can be expressed as a sum of two components:

$$q_\Sigma(t) = q_o(t) + q(t) ,$$

where $q_o(t)$ is the oscillation produced directly by the pumping [which can be computed by setting $x \equiv 0$ and $V_{\text{fluct}} \equiv 0$ into (10.22)], and $q(t)$ is the sum of the responses to the movement $x(t)$ of the mechanical object and to the fluctuating e.m.f. $V_{\text{fluct}}(t)$. Of course, the first term, $q_o(t)$, will be far larger than the second, $q(t)$. It is important to emphasize that the first term is determined uniquely by the pumping, and thus its contribution to the transducer's output can be calculated precisely (and subtracted). Because of this, and because all the information about $x(t)$ is contained in the second term, we shall ignore the first and compute only the second.

The second term, $q(t)$, obeys the equation of motion

$$L\frac{d^2q}{dt^2} + R\frac{dq}{dt} + \frac{q}{d} - \frac{x(t)}{C}\frac{q_o(t) + q(t)}{C} = V_{\text{fluct}}(t) \ .$$

Taking account of the inequality (10.4b) (or the equivalent condition $q_o \gg q$), we can delete the second-order term qx/C, thereby bringing the equation of motion into the form

$$L\frac{d^2q}{dt^2} + R\frac{dq}{dt} + \frac{q}{C} = E(t){\cdot}x(t) + V_{\text{fluct}}(t) \ , \tag{10.23}$$

where

$$E(t) = \frac{q_o(t)}{Cd}$$

is the pumped electric field in the capacitor. From arguments given above it follows that, if the inequalities $1/\tau_M \ll \omega_e$ and $\delta x/d \ll d$ [equations (10.4)] are satisfied, then the parametric action of the mechanical object on the transducer's oscillations (via modulation of the capacitance) is fully described by the e.m.f. $E(t){\cdot}x(t)$.

Maintaining these same assumptions, let us analyze the back action of the capacity transducer on the measured object. The ponderomotive force by which the capacitor plates attract each other is

$$F(t) = \frac{q_\Sigma^2(t)}{2Cd} = \frac{1}{2Cd}[q_o^2(t) + 2q_o(t){\cdot}q(t) + q^2(t)] \ .$$

The first term, $q_o^2/2Cd$, is a definite, predictable force that can be taken into account and compensated (to within the accuracy of our knowledge of the pumping e.m.f.) The third term $q^2/2Cd$ is of second order (because $q \ll q_o$) and thus can be discarded. Therefore, the ponderomotive force can be regarded as equal to

$$F(t) = \frac{q_o(t)q(t)}{2Cd} = E(t)q(t) \ . \tag{10.24}$$

Equations (10.23) and (10.24) completely describe the behavior of the capacity transducer. Notice that the variables $x(t)$ and $q(t)$ appear in them only at first order, i.e. the equations are linear. This linearization, which is made possible by condition (10.4b), simplifies substantially the analysis of measuring devices based on capacity transducers.

An important feature of a capacity transducer is the fact that, because of the high quality factor of its resonator (i.e., its circuit), its oscillations are almost precisely sinusoidal. This permits it to be analyzed using the "slowly-varying-amplitude" approximation.[49] This approximation is based on the change of variable

$$q(t) = q_c(t)\cos\omega_p t + q_s(t)\sin\omega_p t \ ,$$

$$\frac{dq(t)}{dt} = -\omega_p q_c(t)\sin\omega_p t + \omega_p q_s(t)\cos\omega_p t \ ,$$

where ω_p is the pump frequency, which shows up in the transducer's equations of motion as the frequency of $E(t)$,

$$E(t) = E_o(t)\cos\omega_p t \ ,$$

and which (we recall) was chosen to be very close to the resonator's eigenfrequency

$$|\omega_e - \omega_p| \ll \omega_e \ .$$

The new variables $E_o(t)$, $q_c(t)$, and $q_s(t)$ vary on timescales much longer than the resonator and pump periods. By inserting these expressions into (10.23) and discarding negligibly small terms, we obtain

$$\frac{dq_c(t)}{dt} + \frac{1}{\tau_e^*}q_c(t) + \Gamma q_s(t) = -C\omega_e V_s(t) \ ,$$

$$\frac{dq_s(t)}{dt} + \frac{1}{\tau_e^*}q_s(t) - \Gamma q_c(t) = \frac{C\omega_e}{2}E_o(t)x(t) + C\omega_e V_c(t) \ , \quad (10.25)$$

where

$$\Gamma = \frac{\omega_p^2 - \omega_e^2}{2\omega_p} \simeq \omega_p - \omega_e \ ,$$

and where

$$V_s(t) = 2\overline{V_{\text{fluct}}(t)\cdot\sin\omega_e t} \ , \quad V_c(t) = 2\overline{V_{\text{fluct}}(t)\cdot\cos\omega_e t}$$

are slowly changing noise e.m.f.'s. (The averaging in these noise e.m.f.'s is over the period $2\pi/\omega_p$ of the high-frequency oscillations.) These same approximations bring equation (10.24) into the form

$$F(t) = \frac{1}{2}E_o(t)\cdot q_c(t) \ . \quad (10.26)$$

Equations (10.25) and (10.26) will be the foundation for analyzing two different regimes of nonstationary operation, in the next two sections.

In concluding this section, let us note that equations (10.23)—(10.26) can be used in fully quantum mechanical calculations as well as classical and semiclassical ones. To make them fully quantum mechanical, one need only replace the classical $x(t)$, $q(t)$, $q_c(t)$, and $q_s(t)$ by their corresponding Heisenberg-picture operators. Since these equations are linear, failure of the operators to commute has no influence on the form of the equations or its solutions. [This is evidently true only in the first step of the indirect measurement: the dynamical interaction of the object with the transducer's probe oscillator. In the second step, the extraction of the information from the probe (i.e. from the transducer's circuit), non-commutivity is important, as usual.]

10.6* Frequency upconverter

Let us assume that the pumping is perfectly sinusoidal, and the pump frequency is equal to the difference between the eigenfrequencies of the transducer's circuit and the mechanical oscillator:

$$E(t) = E_o \cos(\omega_e - \omega_M)t \ .$$

Then the input signal to the capacity transducer is, according to equation (10.23),

$$E(t)x(t) = E_o \cos(\omega_e - \omega_M)t \cdot x_o \sin(\omega_M t + \phi_M)$$

$$= \frac{E_o x_o}{2} \{ \sin(\omega_e t + \phi_M) - \sin[(\omega_e - 2\omega_M)t - \phi_M] \} \ .$$

Here x_o and ϕ_M are the amplitude and phase of the mechanical oscillations. If the circuit's quality factor is high enough, $\omega_M \tau_e^* \gg 1$, then the circuit will respond far more weakly at the frequency $\omega_e - 2\omega_M$ than at the resonant frequency ω_e; cf. Fig. 10.4. Therefore, we can regard

$$\frac{E_o x_o}{2} \sin(\omega_e t + \phi_M)$$

as the only e.m.f. exciting the circuit.

The back action force that the transducer exerts on the mechanical oscillator has the form [cf. equation (10.24)]

$$F(t) = E_o \cos(\omega_e - \omega_M)t \cdot A \sin(\omega_e t + \phi_e)$$

$$= \frac{E_o A}{2} \{ \sin(\omega_M t + \phi_e) + \sin[(2\omega_e - \omega_M)'t - \phi_M] \}.$$

Here A and ϕ_e are the amplitude and phase of the oscillating charge on

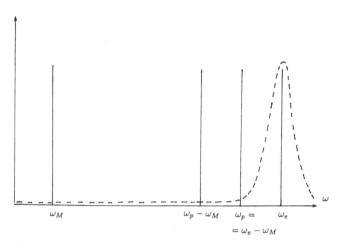

Fig. 10.4 The spectral components of the oscillations of a capacity transducer's circuit when it is operating as a frequency upconverter.

the circuit's capacitor plates. The second (high-frequency) component hardly affects the mechanical oscillator at all, and thus can be ignored, leaving

$$F(t) = \frac{E_o A}{2} \sin(\omega_M t + \phi_e) .$$

The above equations show that both oscillators (high-frequency electrical and low-frequency mechanical) "see" each other's oscillations as occurring at their own frequencies. The system thus behaves like a pair of identical coupled oscillators. The dynamics of such a system is characterized by beatings—periodic interchanges of oscillation energy between the two oscillators. The beat period is determined by the strength of the coupling, in this case by the amplitude of the pump electric field E_o. The beat period can be computed using the equations of motion (10.25) for the electrical oscillator, and the equation of motion for the mechanical oscillator driven by the force (10.26):

$$\frac{dq_c(t)}{dt} + \frac{1}{\tau_e^*} q_c(t) - \omega_M q_s(t) = -C \omega_e V_s(t) ,$$

$$\frac{dq_s(t)}{dt} + \frac{1}{\tau_e^*} q_s(t) - \omega_M q_c(t) = \frac{C \omega_e}{2} E_o(t) x(t) + C \omega_e V_{s(t)} , \quad (10.27)$$

$$\frac{d^2 x(t)}{dt^2} + \omega_M^2(t) = \frac{E_o}{2m} q_c(t) .$$

Here we have used $\Gamma = -\omega_M$, a relation valid for our upconverter situation. The dynamics of this coupled capacity-transducer-plus-mechanical-oscillator system is determined by its set of eigenfrequencies, which are the solutions to the characteristic equation that follows from (10.27):

$$\det \begin{bmatrix} i\omega + \dfrac{1}{\tau_e^*} & -\omega_M & 0 \\[2mm] \omega_M & i\omega + \dfrac{1}{\tau_e^*} & -\dfrac{C\,\omega_e E_o}{2} \\[2mm] \dfrac{-E_o}{2m} & 0 & -\omega^2 + \omega_M^2 \end{bmatrix} = 0 \ .$$

If the electrical resonator's relaxation time is far longer than the mechanical oscillator's period ($\omega_M \tau_e^* \gg 1$), then the eigenfrequencies are

$$\omega_{1,2} = \left[\omega_M^2 \pm \sqrt{\dfrac{C\,\omega_e \omega_M E_o^2}{4m}} \right]^{1/2} \simeq \omega_M \pm \dfrac{1}{4}\sqrt{\dfrac{C\,\omega_e E_o^2}{m\,\omega_M}} \ .$$

The difference between these eigenfrequencies determines the beat period:

$$\tau_B = \dfrac{4\pi}{\omega_1 - \omega_2} = \dfrac{8\pi}{E_o}\sqrt{\dfrac{m}{C}} \ . \tag{10.28}$$

The nature of the information about the mechanical oscillator that the observer obtains from this measurement scheme is determined completely by the observer's procedure for extracting information from the transducer's circuit. In particular, when the circuit's generalized coordinate (the charge on its capacitor) is monitored continuously, e.g. using any linear amplifier, then the observer in effect monitors the mechanical oscillator's amplitude, but with a time delay of half the beat period. The sensitivity to an external force on the mechanical oscillator, in this case, is clearly constrained by the standard quantum limit.

Any other, more sophisticated scheme for monitoring the transducer's circuit is equivalent to applying some analogous scheme directly to the mechanical oscillator (though with some limitations that will be discussed below). For example, devices for monitoring the quadrature amplitudes of an electrical signal ("parametric amplifiers") have long been known in electronics, and in recent years their sensitivities have approached the quantum level. By coupling such a parametric amplifier to the transducer's circuit, one can achieve a measurement of the mechanical oscillator's quadrature amplitude.

A similar approach can be used for a QND measurement of the number of quanta in the mechanical oscillator: QND measurements of electromagnetic quanta are likely to be achieved within the next few years, but it looks far more difficult to measure directly the number of quanta in a mechanical oscillator. There are two reasons. First, the energy of mechanical quanta is far smaller, and second, the practical realization of the required interaction of a mechanical oscillator with the measuring device is far more complex. Frequency upconversion may solve both problems, at least in principle. The beatings convert each quantum of mechanical oscillation energy $\hbar\omega_M$ into a much more energetic electromagnetic quantum $\hbar\omega_e$. The number of electromagnetic quanta can be registered by a QND energy-measuring device, without interfering with their transformation back into mechanical quanta half a beat period later.

In conclusion, we must note that such parametric upconversion devices can be realized in practice only if the level of dissipation is sufficiently low. The relaxation times τ^* of both the mechanical and the electrical oscillators must be substantially longer than the beat period—which has a value, for example, of one second when $m = 10^4$ g, $C = 1$ cm, $E_o = 10^6$ volt/cm.

10.7* Capacity transducer with two-side-band pumping

At the end of section 10.2 we noted that one can operate a capacity transducer in a regime that permits one to measure a quadrature amplitude of a mechanical oscillator with an accuracy better than the standard quantum limit. We shall now analyze the details of this regime.

This regime is characterized by a pump field that varies as

$$E(t) = E_o \cos\omega_e t \cdot \cos\omega_M t \ . \tag{10.29}$$

This time variation can also be expressed as a sum of two sinusoidal components with frequencies that are the two side bands $\omega_e - \omega_M$ and $\omega_e + \omega_M$ of the transducer's eigenfrequency:

$$E(t) = \frac{E_o}{2}[\cos(\omega_e - \omega_M)t - \cos(\omega_e + \omega_M)t] \ .$$

For this reason, one calls this "two-side-band pumping." The transducer's input signal in this case is proportional to the product of the mechanical coordinate $x(t)$ and $\cos\omega_M t$, as one sees from the schematic diagram in Fig. 8.3:

$$E(t) \cdot x(t) = E_o \cos\omega_e t \cos\omega_M t [X_1 \cos\omega_M t + X_2 \sin\omega_M t] \ ,$$

where X_1, X_2 are the mechanical oscillator's quadrature amplitudes. The inertial averaging device shown in Fig. 8.3 is here the transducer's circuit with its high quality factor, and this averaging smooths out any changes of the input amplitude that occur on timescales smaller than τ_e^*.

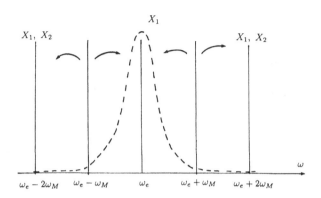

Fig. 10.5 The spectral components of the oscillations of a capacity transducer's circuit with two-side-band pumping.

The electrical equations of motion (10.25) for this case have the form (obtained by setting $E_o(t) = E_o \cos\omega_M t$ and $\Gamma = 0$):

$$\frac{dq_c(t)}{dt} + \frac{1}{\tau_e^*} q_c(t) = -C\omega_e V_s(t) ,$$

$$\frac{dq_s(t)}{dt} + \frac{1}{\tau_e^*} q_s(t) \qquad\qquad (10.30)$$

$$= \frac{C\omega_e}{2} E_o \cos\omega_M (X_1 \cos\omega_M t + X_2 \sin\omega_M t) + C\omega_e V_c(t) .$$

In the useful output signal, as in the input (first term on the right-hand-side of the second equation), there are three spectral components: the central resonance and two sidebands, detuned from the central resonance by $2\omega_M$. These output frequencies result from the interaction of the two-side-band pumping (10.21) with the low-frequency mechanical oscillations [cf. Fig. 10.5]:

$$\cos\omega_M t (X_1 \cos\omega_M t + X_2 \sin\omega_M t) \equiv \frac{X_1}{2} + \frac{1}{2}(X_1 \cos2\omega_M t + X_2 \sin2\omega_M t) .$$

As should be evident from equations (10.30), if the transducer's circuit has a high enough quality factor ($\omega_M \tau_e^* \gg 1$), then the response to the central component will be approximately $\omega_M \tau_e^*$ larger than the response to the two side bands (which are outside the circuit's resonance). The most important fact here is that the central component depends only on the quadrature amplitude X_1, and therefore the strength of the information about X_2 is reduced relative to it by a factor $\omega_M \tau_e^*$.

Thus, this scheme approximately satisfies the main criterion for a QND measurement: information about the undesired quadrature component X_2 is suppressed, but not completely. The level of suppression is determined by the quality factor of the transducer's circuit. Because of this, the precision with which the desired component X_1 is measured can be better than the standard quantum limit, but it cannot be arbitrarily high.

Let us assume that the information is extracted from the capacity transducer using an "ordinary" detector, i.e. one that monitors the transducer's oscillations without any additional filtering. Then the ratio of the errors for X_1 and X_2 will be

$$\frac{\Delta X_1}{\Delta X_2} = \frac{1}{\omega_M \tau_e^*} \ .$$

On the other hand, their product must satisfy the uncertainty relation

$$\Delta X_1 \cdot \Delta X_2 \geq \frac{\hbar}{2m \omega_M} \ .$$

If there are no excess noises in the system, beyond the pure quantum ones, then this relation will be an equality, and when combined with the preceding relation, it will give

$$\Delta X_1 = \sqrt{\frac{\hbar}{2m \omega_M^2 \tau_e^*}} \ , \qquad \Delta X_2 = \sqrt{\frac{\hbar \tau_e^*}{2m}} \ . \tag{10.31}$$

Therefore, the precision with which X_1 is measured in this regime of two-side-band pumping can be $\sqrt{\omega_M \tau_e^*}$ times better than the standard quantum limit. Correspondingly, if one is using this device to detect the action of an external force on the mechanical oscillator, then the sensitivity to the force will also be a factor $\sqrt{\omega_M \tau_e^*}$ better than the standard quantum limit:

$$F_{min} = \frac{1}{\xi \tau_F} \sqrt{\frac{\hbar m}{2 \tau_e^*}} \ . \tag{10.32}$$

XI   Quantum nondemolition measurements of a resonator's energy

11.1 Review of methods of measurement

There are two different variants of the task of measuring electromagnetic energy: i) measuring the energy in a mode of an electromagnetic resonator, and ii) measuring the energy of a traveling electromagnetic wave. This chapter deals with the first variant; the next chapter, with the second.

The general principles underlying QND measurements of energy were formulated in chapter IV. Let us recall that the main principles are: i) the response of the measuring device must be directly proportional to the energy (and not, for example, to the strength of the electric field or the charge on the capacitor); and ii) the response must contain no information about the phase of the electromagnetic oscillations. Chapter IV analyzed the simplest example of such a device: a ponderomotive sensor that registers the electromagnetic pressure on the resonator's wall.

The weakness of the electromagnetic pressure produced by a small number of quanta makes it unlikely that this ponderomotive method can be realized in practice. Because of this, several other schemes for QND energy measurements have been proposed. Most are based on nonlinear effects in dielectrics. Their practical realization is a rather complicated task because, when the energy density is low, the nonlinear effects hardly work at all, and the response of the measuring device is correspondingly small. To register the device's response, one must use a rather long averaging time, and this requires a long lifetime of the electromagnetic quanta in the nonlinear material. However, presently known materials with large nonlinearities have substantial electromagnetic losses, while those with low losses have low nonlinearities.

To increase the influence of the nonlinearities, for a fixed number of electromagnetic quanta, one can decrease the volume that the quanta occupy (and thereby increase their energy density). From this viewpoint, there are good prospects for optical dielectric microresonators[50] in which the volume occupied by the electromagnetic field is very small ($\approx 10^{-10}$ cm³). The high concentration of the energy permits one to use a dielectric material with relatively low nonlinearities, and thus with low losses.

Estimates show that (for presently known dielectric materials), the best results can be obtained using cubic dielectric nonlinearities.[51,52] The numbers suggest that, by concentrating the energy in the microresonator's small volume, it may be possible to achieve a precision of a fraction of an optical quantum.[53] The details of this method will be analyzed in the next section. Here we shall briefly discuss two other methods

1. The inverse Faraday effect[54]

When a magnetoactive material is penetrated by a polarized electromagnetic wave, it becomes magnetized, and its magnetization $\vec{I}$ is proportional to the flux of electromagnetic energy:

$$\vec{I} = \frac{V^*}{\omega}\vec{\Pi} \ .$$

Here $\vec{\Pi}$ is the Poynting vector, ω is the electromagnetic wave's frequency, and V^* is the material's Verdet constant. By measuring the magnetization, one can infer the energy of the electromagnetic field.

Estimates show that to count a few optical quanta, one must use a material that combines the transparency of the best known optical glasses (attenuations of ≈ 0.2 db/km) with the highest known Verdet constants (≈ 1 oersted^{-1} cm^{-1}). No presently known materials have this combination of parameters. (The best known are certain garnets.)

2. Optical detection

In nonlinear optics materials with second-order dielectric nonlinearities are widely used. The dielectric polarization that an electromagnetic wave induces in such a material contains a component proportional to the square of the electric field strength:

$$P^{(2)} = \chi^{(2)}[E_o\cos(\omega t + \phi)]^2 = \frac{1}{2}\chi^{(2)}E_o^2[1 + 2\cos 2(\omega t + \phi)] \ .$$

Here $\chi^{(2)}$ is the coefficient of second-order nonlinear dielectric susceptibility, E_o is the amplitude of the electric field oscillations, ω is the oscillation frequency (presumed to be in the optical band), and ϕ is the phase. The slowly changing part of the polarization $\chi^{(2)}E_o^2/2$, is proportional to the light's energy density and contains no information about its phase. Thus, by monitoring the polarization with a device that averages out the oscillations at the optical frequencies ω and 2ω, one can measure the electromagnetic energy in a QND way.

For example, a piece of nonlinear dielectric could be placed inside a resonator that has an eigenfrequency $\omega_{resonator}$ much lower than ω. Then, when a high-frequency electromagnetic pulse with duration $\tau \simeq \omega_{resonator}^{-1}$ passes through the resonator, it will induce oscillations with amplitude proportional to the pulse's energy. The main obstacle to realizing such a scheme is the same as for the previous one: no known material has simultaneously a sufficiently low level of dissipation and a sufficiently high level of second-order nonlinearity.

11.2 Measuring device based on cubic dielectric nonlinearity

In a dielectric with cubic nonlinearity, the polarization induced by an electromagnetic field has a component proportional to the cube of the electric field strength $E(t)$:

$$P^{(3)} = \chi^{(3)}[E(t)]^3 .$$

here $\chi^{(3)}$ is the third order coefficient of nonlinear dielectric susceptibility. One consequence of this is a dependence of the material's refractive index on the density ρ of electromagnetic energy in it:

$$n = n_o + n_2\rho , \tag{11.1}$$

where

$$n_2 = \frac{12\pi^2\chi^{(3)}}{n_0} .$$

If we place such a dielectric in a resonator, then the refractive index will change by an amount that depends on the resonator's electromagnetic energy. One can measure this change by the phase shift or delay that it produces in a probe electromagnetic wave sent through the dielectric. Methods for measuring tiny phase shifts and delays are very well developed. As a result (at least for presently known nonlinear dielectrics), this method promises a higher precision for QND energy measurements than those discussed in the last section.

A possible scheme for such a measuring device is shown in Fig. 11.1. We denote by V the volume in which the measured energy E is located. This volume might, for example, be the interior of a Fabry-Perot

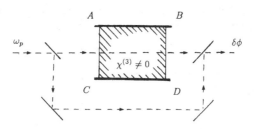

Fig. 11.1 A scheme for a QND measurement of electromagnetic energy in a nonlinear resonator.

resonator with eigenfrequency ω and with mirrors in the planes AB and CD of the figure. This volume is filled with a nonlinear dielectric whose index of refraction has the form (11.1). A probe electromagnetic wave with frequency ω_p passes through it in the x-direction, and after it leaves the dielectric, its phase is measured (e.g., by interferometry).

In this section, for simplicity, we assume that there is no absorption. The influence of absorption will be analyzed in the next section.

It is not difficult to show, using equation (11.1), that the speed of the probe wave in the dielectric is

$$v = \frac{c - n_2 LW/V}{n_0 + n_2 \boldsymbol{E}/V} \ ,$$

where c is the speed of light in vacuum, L is the length of the dielectric in the x direction, and W is the power in the probe wave. Thus, the change of phase of the probe wave as it passes through the dielectric is

$$\delta\phi = \omega_p \frac{L}{v} \simeq \omega_p \tau_o \left[1 + \frac{n_2}{V n_o}(\tau_o W + \boldsymbol{E}) \right] \ , \tag{11.2}$$

where $\tau_o = L n_o/c$.

This signal phase shift (the part of $\delta\phi$ proportional to $\boldsymbol{E}$) must be measured amidst a background of phase fluctuations. Assuming that the coherence time of the probe wave is much longer than the measurement time, and assuming no excess noise due to practical problems, the phase fluctuations are described by the following formula (cf. section 12.1):

$$S_\phi = \frac{\hbar\omega_p}{4\langle W \rangle} \ ,$$

where S_ϕ is the spectral density of the phase fluctuations and $\langle W \rangle$ is the mean power in the probe wave.

The power W is not precisely constant; rather, it is characterized by shot noise. The spectral density of the shot noise is

$$S_W = \hbar\omega_p <W> .$$

Equation (11.2) implies that this shot noise produces an additional random phase shift of the probe wave

$$\delta\phi_{\text{self action}} = \frac{\omega_p \tau_o^2}{Vn_0}(W - <W>)n_2 ,$$

and a corresponding additional error in the measurement. This effect is sometimes called the "self-action" of the probe wave on its own phase.

Suppose that the measurement time (the time over which the phase is averaged) is equal to τ_{measure}. Then the error ΔE in the energy measurement is given by

$$(\Delta E)^2 = \frac{1}{\tau_{\text{measure}}}\left[\frac{Vn_0}{\omega_p\tau_o n_2}\right]^2 \cdot \left[S_\phi + \left[\frac{\omega_p\tau_o^2 n_2}{Vn_0}\right]^2 S_W\right]$$

$$= \frac{\hbar\omega_p}{\tau_{\text{measure}}}\left[\frac{1}{4<W>}\left[\frac{Vn_0}{\omega_p\tau_o n_2}\right]^2 + \tau_o^2<W>\right] .$$

The error $\Delta N = \Delta E/\hbar\omega$ in the number of quanta corresponding to this energy error is

$$(\Delta N)^2 = \frac{1}{\tau_{\text{measure}}}\left[\frac{(N^*)^2}{4\omega^2 N_p\tau_o} + \left[\frac{\omega_p}{\omega}\right]^2 N_p\tau_o\right] ,$$

where N_p is the mean number of probe quanta in the dielectric,

$$N_p = \frac{<W>\cdot\tau_o}{\hbar\omega_p} ,$$

and

$$N^* = \frac{Vn_o}{\hbar\omega n_2}$$

is a useful parameter for characterizing the resonator's nonlinearity. It is easy to show that $1/N^*$ is the fractional shift of the resonator's eigenfrequency when its energy is changed by one quantum.

The higher is the power of the probe wave, the lower are the shot-noise-induced phase fluctuations, but the stronger is the self action. There is an optimal power at which the contributions of the two to the total measurement error are equal. The number of probe quanta corresponding to this power is

$$N_p^{\text{optimum}} = \frac{N^* \omega}{2\omega_p^2 \tau_0} .$$

In this optimal case, the number of quanta in the resonator can be measured with an accuracy up to

$$\Delta N = \sqrt{\frac{N^*}{\omega \tau_{\text{measure}}}} . \tag{11.3}$$

In conclusion, it is important to note that the influence of the self action can be removed, in principle, either by sending the probe wave subsequently through a dielectric with the opposite sign of n_2, or by using a dielectric with different levels of nonlinearity n_2 at the probe and signal frequencies. If the self action is removed, then the second term in equation (11.3) disappears, and the measurement error becomes

$$\Delta N = \frac{N^*}{2\omega_p \sqrt{N_p \tau_o \tau_{\text{measure}}}} . \tag{11.4}$$

This error will be smaller, the higher is the power of the probe wave.

11.3 The role of dissipation

At optical frequencies the temperature of the measuring device is generally far below the quantum level,

$$k_B T \ll \hbar \omega$$

In this case dissipation can be described using a simple and completely adequate model based on classical probability theory.

Each quantum in the dielectric has a probability of surviving for a time τ given by

$$P(\tau) = e^{-\tau/\tau^*} ,$$

where τ^* is the relaxation time. (For the best optical fibers now available, this τ^* is of order 10^{-3} sec.) The probability of the photon being absorbed, correspondingly, is

$$1 - P(\tau) = 1 - e^{-\tau/\tau^*} .$$

The probability that, beginning with an initial number of quanta N, a number N_1 remain after time τ is described by the binomial distribution:

$$w(N_1 | N) = \frac{N!}{N_1!(N-N_1)!} [P(\tau)]^{N_1} [1-P(\tau)]^{N-N_1} .$$

The mean number of surviving photons is

$$\langle N_1 \rangle = \sum_{N_1=0}^{N} N_1 w(N_1|N) = Ne^{-\tau/\tau^*} ,$$

and the variance is

$$(\Delta N_1)^2 = \sum_{N_1=0}^{N} N_1^2 w(N_1|N) - \langle N_1 \rangle^2 = Ne^{-\tau/\tau^*}(1-e^{-\tau/\tau^*}) .$$

In high precision experiments, the measuring time is generally much less than the relaxation time. Taking this into account, we can rewrite these two formulae with the exponentials expanded as series. Doing so, we find for the mean number of quanta absorbed during time τ

$$N - \langle N_1 \rangle \simeq N\frac{\tau}{\tau^*} ,$$

and for the rms uncertainty in the remaining number of quanta

$$\Delta N_1 = \sqrt{N\frac{\tau}{\tau^*}} . \tag{11.5}$$

In the measurement scheme of the last section, noise can arise from both absorption of the resonator quanta being measured, and absorption of the probe wave's quanta. However, these two types of absorption affect the final measurement accuracy in very different ways. The probe wave enters the resonator in a nearly coherent state. In a coherent state, however, the uncertainties in the energy and phase are substantial: approximately $\sqrt{\tau^*/\tau_o}$ larger than the uncertainties introduced by dissipation. Thus, when $\tau_o \ll \tau^*$, dissipation of the probe wave can be neglected. (This may change if the probe wave begins in a non-coherent state, for example, in a state with squeezed phase; however, we shall not discuss this possibility here.) By contrast with the probe wave, absorption of the resonator's quanta can produce noise ΔN_1, that is substantial if one is trying to measure the number of quanta with an error much less than $\sqrt{N}$.

Equations (11.3)—(11.5) show that the measurement precision is higher, the longer is the total duration of the measurement, $\tau = \tau_o + \tau_{measure}$. But the longer is τ, the larger is the probability that the uncertainty ΔN_1 induced by dissipation will exceed the measurement error. When it does so, the "exactly measured" number will be some not very interesting random number that occurs at some moment during the measurement.

The optimal measurement time is that which makes the errors (11.3) and (11.5) equal:

$$\tau_{measure} = \sqrt{\frac{N^*\tau^*}{N\omega}} , \tag{11.6}$$

and the corresponding optimal sum of their errors is

$$\Delta N = \left[N\frac{\tau_o}{\tau^*} + 2\sqrt{\frac{NN^*}{Q}} \right]^{\frac{1}{2}},$$

where $Q - \omega\tau^*$ is the resonator's quality factor. It is desirable, according to this formula, for the probe wave to spend a total time in the resonator that is much less than the measurement time: $\tau_o \ll \tau_{\text{measure}}$. Then the measurement error is

$$\Delta N = \left[\frac{4NN^*}{Q} \right]^{1/4}. \tag{11.7}$$

Thus, to beat the standard quantum limit $\Delta N_{\text{SQL}} = \sqrt{N}$, it is necessary to have

$$4\frac{N^*}{Q} < N, \tag{11.8}$$

and to reach a precision of one quantum requires

$$4\frac{N^*}{Q} < \frac{1}{N}. \tag{11.9}$$

In the absence of self-action, one must minimize the sum of the errors (11.3) and (11.4). The minimum occurs for

$$\tau_o = \tau_{\text{measure}} = \left[\frac{(N^*)^2\tau^*}{4NN_p\,\omega_p^2} \right]^{1/3}, \tag{11.10}$$

and the minimum total error is

$$\Delta N = \sqrt{3}\left[\frac{NN^*}{2\sqrt{N_p}\,\omega_p\tau^*} \right]^{1/3}. \tag{11.11}$$

11.4* Resonator coupled to a waveguide

The previous sections ignored the problem of how to insert the measured energy into the nonlinear resonator. When the insertion is done via a waveguide, from the waveguide's viewpoint the resonator is an inhomogeneity that reflects a portion of the wave's energy, and this leads to an additional measurement error. This problem does not arise if the measurement is performed directly on the traveling wave. However, as we have discussed, the accumulation of energy in the resonator's tiny volume is needed in order to achieve sufficient nonlinearity for the measurement (at least for presently known dielectrics and the present state-of-the-art).

One could achieve perfect matching of the waveguide into the resonator if the resonator's eigenfrequency were precisely fixed and the incoming wave had a sufficiently small bandwidth

$$\delta\omega \ll \frac{1}{\tau^*_{\text{loaded}}} . \tag{11.12}$$

Here τ^*_{loaded} is the "loaded" relaxation time, determined by the resonator's coupling to the waveguide. (We assume that the "free" relaxation time τ^* associated with the resonator's dissipation is far longer than the loaded relaxation time.) However, the energy-phase uncertainty relation for the resonator during the measurement dictates that the relative frequency of the resonator and the wave that drives it can be known only with an accuracy

$$\delta\omega \geq \frac{\hbar\omega_e}{2\tau_{\text{measure}}\Delta E} ,$$

where ΔE is the measurement error. To register the signal it is necessary that $\tau_{\text{measure}} < \tau^*_{\text{loaded}}$. Therefore, if one attempts to measure the energy with the accuracy of one quantum, $\Delta E \leq \hbar\omega$, the waveguide and resonator will be unmatched,

$$\delta\omega \geq \frac{1}{\tau^*_{\text{loaded}}} .$$

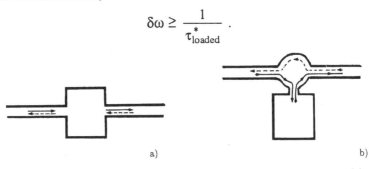

a)   b)

Fig. 11.2 Two schemes for coupling a waveguide to a resonator: (a) a "fly-through" scheme; (b) a "reflection" scheme. The continuous line is the signal wave; the dashed line is the zero-point oscillation of empty modes.

Let us analyze the measurement scheme shown in Fig. 11.2a. The resonator is inserted into a break in the waveguide. The signal propagates from left to right, and zero-point oscillations propagate in the opposite direction. To the resonator is coupled a measuring device (not shown in the picture), and in accord with the general theory of QND measurements, the Hamiltonian of that coupling is proportional to the resonator's energy operator $\hat{E}$. Thus, the Hamiltonian for the entire system has the form

$$\hat{H} = \hat{E}(1-\hat{X}) + \hat{H}_e \ , \tag{11.13}$$

where $\hat{H}_e$ is the Hamiltonian for the waveguide and its coupling to the resonator, and $\hat{X}$ is a dimensionless generalized coordinate of the measuring device. (For the device analyzed in section 11.2, $\hat{X}$ is proportional to the energy of the probe wave.) The signal going into the measuring device is a change in the generalized momentum $\hat{P}$ that is canonically conjugate to $\hat{X}$.

We shall assume that the quantity measured is the energy $\hat{E}$ of a narrow-band wave packet [one satisfying condition (11.12)], with frequency ω_e. This energy is uniquely related to the wave packet's number operator $\hat{N}$ by $\hat{E} = \hbar\omega_e\hat{N}$. In the Heisenberg picture, the resonator's energy operator evolves in the following manner:

$$\hat{E}(t) = \frac{\hbar}{\pi\tau_{\text{loaded}}^*} \int_0^\infty \left[\frac{\hat{\omega}_x^2}{\omega\omega'} + 1 \right] \hat{A}^\dagger(\omega)\hat{A}(\omega')e^{i(\omega'-\omega)t} \frac{d\omega d\omega'}{\sqrt{\omega\omega'}} \tag{11.14}$$

plus a rapidly oscillating component. Here

$$\hat{A}(\omega) = \frac{\hat{a}_1(\omega)+\hat{a}_2(\omega)}{\hat{Z}(\omega)} \ ,$$

$$\hat{\omega} = \omega_e(1-\hat{X}) \ ,$$

$$\hat{Z}(\omega) = \frac{\hat{\omega}_x^2}{\omega^2} - 1 - \frac{2i}{\omega\tau_{\text{loaded}}^*} \ ,$$

and $\hat{a}_{1,2}^\dagger$, $\hat{a}_{1,2}(\omega)$ are the creation and annihilation operators for leftward and rightward propagating modes; see Fig. 11.2.

The final value of the operator $\hat{P}$ at the end of the measurement, in the approximation $\omega_e\tau_{\text{loaded}}^* \gg 1$ is

$$\hat{P}_{\text{final}} = \hat{P} + \int_0^\infty \hat{E}(t)dt = \hat{P} + \hbar\omega_e\tau_{\text{loaded}}^* \int_0^\infty \hat{A}^\dagger(\omega)\hat{A}(\omega)d\omega \ ;$$

cf. equation (11.14). The state of a narrowband, N-quantum wave packet propagating rightward in the waveguide toward the resonator is (see the next chapter)

$$|N> = \int_0^\infty \phi(\omega_1,...,\omega_N)\hat{a}_1^\dagger(\omega_1) \cdots \hat{a}_1^\dagger(\omega_N)d\omega_1 \cdots d\omega_N|0> \ ,$$

where

$$|\phi(\omega_1,...,\omega_N)|^2 \to \delta(\omega_1-\omega_e) \cdots \delta(\omega_N-\omega_e) \ ,$$

and $|0>$ is the waveguide's vacuum state.

One can show (we omit all the rather long, intermediate calculations) that the characteristic function of the operator $\hat{P}_{final}$,

$$\chi(q) = <\exp(-i\hat{P}_{final}q)>$$

is equal to

$$\chi(q) = \text{Tr}[\exp(-i\hat{P}q)\hat{\rho}] ,$$

where

$$\hat{\rho} = \sum_{n=0}^{N} w(n|N)\hat{U}^n |\psi><\psi|(\hat{U}^\dagger)^n , \qquad (11.15)$$

$|\psi>$ is the measuring device's initial state, and

$$w(n|N) = \frac{N!}{2^N n!(N-n)!} , \qquad \hat{U} = \frac{\hat{X}-\dfrac{i}{\omega_e \tau_{loaded}^*}}{\hat{X}+\dfrac{i}{\omega_e \tau_{loaded}^*}} . \qquad (11.16)$$

From these relations it follows (taking account of the fact that the operator $\hat{X}$ canonically conjugate to $\hat{P}$ is an integral of the motion), that in the Schrödinger picture the measuring device's final state is described by the density operator (11.15)

If all the steps in the measurement procedure are performed precisely, and if the initial wave function is chosen in an optimal way, then the statistics of the possible results of the measurement are described by the distribution (11.16), and the resulting rms measurement error is

$$\Delta N = \sqrt{N} .$$

Thus, although the measuring device permits one to measure precisely the number of quanta *inside* the resonator, the precision of measurement of (fixed) number of quanta passing *through* the resonator is constrained by the standard quantum limit.

This scheme fails to achieve the precision inherent in the QND procedure because of the uncertainty in the time that the quanta spend inside the resonator—an uncertainty produced by the random character of their reflection from the resonator's input and output. It is known that the cause of the stochasticity in the reflection is that zero-point fluctuations of empty modes get superposed on the signal that drives the resonator. To avoid this, one must use an alternative scheme of coupling the resonator to the waveguide: a scheme in which no other modes get superposed on the signal mode.

A possible version of such a scheme is shown in Fig. 11.2b. This scheme uses a circulator to force the empty, leftward-propagating modes to bypass the resonator and, at the same time, force the signal mode into the resonator. For this scheme the evolution of the signal mode's energy operator is given by

$$\hat{E}(t) = \frac{2\hbar}{\pi\tau_{\text{loaded}}^*} \int_0^\infty \frac{\frac{\omega_x^2}{\omega\omega'}+1}{Z^*(\omega)Z(\omega')} \, \hat{a}_1^\dagger(\omega)\hat{a}_1(\omega)e^{i(\omega-\omega')t} \frac{d\,\omega d\,\omega'}{\sqrt{\omega\omega'}} \qquad (11.17)$$

plus rapidly varying components. The final value of the momentum in the measuring device in this case is

$$\hat{P}_{\text{final}} = \hat{P} + 2\hbar\omega_e\,\tau_{\text{loaded}}^* \int_0^\infty \frac{\hat{a}_1^\dagger(\omega)\hat{a}_1(\omega)}{|Z(\omega)|^2} d\,\omega \ .$$

The characteristic function for this operator is

$$\chi(q) = \langle\psi| (\hat{U}^\dagger)^N e^{-i\hat{P}q} \hat{U}^N |\psi\rangle \ .$$

Therefore, after its interaction with the N-quanta, narrowband wave packet, the measuring device will have evolved from its initial state $|\psi\rangle$ to the final state

$$|\psi_N\rangle = \hat{U}^N|\psi\rangle \ . \qquad (11.18)$$

By contrast with the state (11.15), this is a pure state, thereby demonstrating the absence of any additional uncertainties.

If the wave function $|\psi\rangle$ is chosen in such a way that the states of type (11.18) for different values of N are all orthogonal, then the number of quanta can be measured exactly. This can be done by designing the measuring device to measure the operator

$$\zeta = \frac{1}{2\omega_e\tau_{\text{loaded}}^*} \left[\hat{P} + \frac{(\omega_e\,\tau_{\text{loaded}}^*)^2}{2}(\hat{P}\hat{X}^2+\hat{X}^2\hat{P}) \right] \ . \qquad (11.19)$$

The structure of the operator (11.19) reflects the fact that the fraction of the wave-packet energy that penetrates into the resonator is governed by the difference between the wave packet's frequency and that of the resonator (in other words, by the value of X). The term in (11.19) involving $\hat{X}$ corrects for this dependence, thereby permitting the measurement to be exact.

XII   Nonclassical states of electromagnetic waves as tools for quantum measurements

12.1 Quantum properties of a traveling electromagnetic wave

In the mid 1950s, the construction of photodetectors with quantum efficiencies close to unity gave experimenters the possibility of observing in detail the "seed" structure of light. In 1956, one such observation (the Hanbury-Brown-Twiss experiment[55]) demonstrated that in perfectly thermal radiation photons have a tendency to come in pairs.

At about the same time, the invention of quantum auto-oscillators (masers and lasers) stimulated theoretical analyses of the quantum properties of electromagnetic radiation. In particular, Glauber created a complete and consistent theory of the quantum state whose properties are the closest possible to being classical: the so-called coherent state.[56]

When some mode of an electromagnetic resonator is in the coherent state, its oscillations can be expressed as a sum of purely classical oscillations (with precisely defined amplitude and phase), and quantum fluctuations with a constant rms amplitude $\sqrt{\hbar/\rho}$ that is the same as for the zero-point oscillations of the mode's ground state. (Here ρ is the mode's impedance.)

The coherent state can be generated by applying a classical e.m.f. to the mode's ground state. The coherent state is also generated when the mode's generalized coordinate (e.g. the charge on a capacitor) is monitored continuously and with the best possible, constant precision. The uncertainties of both of the mode's quadrature amplitudes in this case are the same and are equal to $\sqrt{\hbar/2\rho}$, and the uncertainty in the number of quanta in the mode is $\sqrt{<n>}$, where $<n>$ is the mean number of quanta; see section 4.1.

One can prepare a traveling electromagnetic wave in a coherent state by, e.g., applying a classical e.m.f. to an electromagnetic resonator that is coupled to a waveguide or transmission line. An "ordinary" classical auto-oscillator that is free of all thermal and other excess noise will emit radiation in the coherent state, and quantum auto-oscillators emit it in a nearly coherent state.

A photodetector, when irradiated by an electromagnetic wave packet in a coherent state, will register a number of photons that cannot be predicted in advance. If the photodetector is ideal (has a quantum efficiency close to unity), then the rms deviation in the number of photons counted (the standard deviation) will be $\sqrt{<n>}$, where $<n>$ is the mean number of photons. Moreover, the statistical properties of any portion of the wave packet, with an arbitrary length $c\tau$, are the same as those of the wave packet as a whole: the standard deviation of the number of registered photons during the time τ is

$$\Delta n = \sqrt{\eta\tau} ,$$

where η is the mean photon flux.

From the classical theory of probability it is known that this is the standard deviation for a collection of randomly emitted particles. The fluctuations are called shot noise, and their strength is characterized by the spectral density

$$S_\eta = \eta . \tag{12.1a}$$

From the number-phase uncertainty relation,

$$\Delta n \cdot \Delta \phi \geq \frac{1}{2} ,$$

and the fact that the coherent state has the minimum fluctuations allowed by the uncertainty principle, one can determine the rms phase fluctuations. In a measurement time τ, they are

$$\Delta \phi = \frac{1}{2\sqrt{\Delta n}} = \frac{1}{2\sqrt{\eta\tau}} ,$$

(and similarly for a resonator containing $\eta\tau$ quanta). This formula can be rewritten in the form

$$\Delta \phi = \sqrt{S_\phi/\tau} ,$$

where

$$S_\phi = \frac{1}{4\eta} \qquad (12.1b)$$

is the spectral density of the phase fluctuations.

In the mid 1970s, theoreticians and experimenters became interested in so-called nonclassical states of the electromagnetic field. This name includes all quantum states except the coherent one, in which thermal and other non-quantum fluctuations are small compared to purely quantum uncertainties. Interest in nonclassical states was aroused by their possible advantages in information transfer[57,58] and in, e.g., parametric transducers for small displacements[59] (cf. chapter X).

There are two classes of especially important nonclassical states for the electromagnetic field. The first, the so-called "amplitude squeezed states," are produced by degenerate parametric amplifiers and generators, and can be observed by phase-sensitive detectors.[60,61] These states are characterized by the fact that the uncertainties of their two quadrature amplitudes $q_{1,2}$ are not the same:

$$\Delta q_1 = \alpha \sqrt{\frac{\hbar}{2\rho}} \,,$$

$$\Delta q_2 = \frac{1}{\alpha} \sqrt{\frac{\hbar}{2\rho}} \,.$$

Here α is the squeeze parameter. It is worth noting that, when one makes a perfect stroboscopic measurement of an oscillator or monitors one of its quadrature amplitudes in a continuous, QND way (chapter VIII), the oscillator is driven into an amplitude-squeezed state.

The second important class of nonclassical states has the following spectral densities for the fluctuations of the number of photons and the phase:

$$S_\eta = \xi\eta \,, \qquad (12.2a)$$

$$S_\phi = \frac{1}{4\xi\eta} \,. \qquad (12.2b)$$

Here ξ is the squeeze parameter (cf. equations (12.1) for a coherent state). States with squeeze parameter $\xi > 1$ are called "phase squeezed" or "photon bunched"; those with $\xi < 1$ are called "energy squeezed" or "photon antibunched." By bunching (grouping) is meant that the photons pass some location not independently, as in the coherent state, but rather in correlated groups. The nonuniformity of the flux (shot noise) in this case is larger than for the coherent state, but the fluctuations of phase are smaller. The opposite effect, antibunching, resembles the suppression of the shot noise in an electrical current in electronics that occurs when the electrons follow each other at more or less constant intervals. In the photon antibunched state, the more orderly is the photon arrival, the greater are the phase fluctuations. In the ideal limit where the number of photons arriving in any chosen time interval is precisely predictable, the phase is completely undetermined.

From electric current with suppressed shot noise, one can obtain light with similar shot-noise suppression. One can do so using a source of light that emits precisely one photon each time an electron passes through it. This method has been demonstrated experimentally using as the source a semiconductor injection laser.[62]

Notice that there is an overlap between the class of amplitude-squeezed states, and the class of energy-squeezed and phase-squeezed states. The amplitude-squeezed state in which

$$\Delta q_1 > \Delta q_2$$

is at the same time phase-squeezed if

$$<q_1> > <q_2> ,$$

or energy-squeezed if

$$<q_1> < <q_2> .$$

However, there are energy-squeezed states that are not amplitude-squeezed.

On the other hand, this classification of nonclassical states is far from complete. There are many quantum states that are not squeezed states. In this chapter we shall analyze an example of such a state: the so-called "frequency-anticorrelated quantum state." This is the optimal state to use in a Doppler measurement of the speed of a mechanical object. Before presenting this analysis, it is appropriate to discuss the main features of QND measurements of the energy of a traveling electromagnetic wave.

12.2 QND measurements of the energy of a traveling electromagnetic wave

In the last chapter it was shown that one can measure with high accuracy the number of quanta and the corresponding energy in a mode of a resonator, using a nonlinear interaction of the mode's field with the probe field of a measuring device. The mode's number of quanta is preserved by such a measurement, and the measuring device strongly perturbs the phase, which is not registered.

It is evident that such a QND measurement is also possible for a traveling wave. One can realize it, at least in principle, using the same nonlinear dielectric effects as for resonators. However, the density of electromagnetic energy in a traveling wave is substantially lower than in a tiny resonator, and therefore a much higher nonlinearity is needed. Estimates show that, using known dielectric materials, it will be very hard to cross the standard quantum limit (though it is worth noting that the energy density can be increased substantially if one uses solitary electromagnetic pulses (solitons), as proposed in Reference 63).

To realize a QND measurement of a traveling wave's energy, another method looks more promising: the interaction of a probe beam of electrons with the wave's electromagnetic field.[64] Estimates show that, at least in principle, this method can register single optical photons ''in flight.'' Let us analyze the main feature of this type of QND measurement:

The experimental setup involves a dielectric waveguide without any outer conducting shield, along which propagates an electromagnetic wave packet of finite duration. An electron travels parallel to the waveguide and very near it, with a speed v_e slightly smaller than the speed of light v_o in the waveguide. During its travel, the electron is given a transverse momentum by the piece of the wave that extends outside the waveguide's dielectric. This transverse momentum δP consists of two components: a ''linear'' one, P_{linear}, which is proportional to the amplitude E_o of the wave's electric field and depends on the wave's phase; and a ''quadratic'' one, $P_{\text{quadratic}}$, which is proportional to E_o^2 and is independent of the phase. The quadratic momentum transfer is caused by the fact that the electron's trajectory, under the action of the linear force, oscillates with the frequency

$$\omega_e = (1 - v_e/v_o)\omega_o \ ,$$

(where ω_o is the electromagnetic frequency), and this oscillation occurs in a spatially nonuniform electric field. The difference in the linear force on the electron when it is nearest the waveguide and farthest away gives rise to a net repulsive, d.c. force (the so-called ''Miller force''[65]):

$$F = \frac{e^2 E_o^2}{2m\,\omega_e^2 x_o} \ .$$

Here x_o is the characteristic lengthscale for the decay of the electric field transverse to the waveguide [$E \sim \exp(-x/x_o)$].

One can show that, if the electron enters the interaction region smoothly and departs smoothly, and if the time of interaction is long enough ($\omega_e \tau \gg 1$), then the total linear momentum transfer will be far smaller than the net quadratic one, $P_{\text{linear}} \ll P_{\text{quadratic}}$. This means that the scattering of the electron by the wave will be proportional to the square of the field amplitude, i.e. to the energy E in the wave

$$\delta P = \frac{E\tau}{d}$$

(where $1/d$ is a coupling constant), and it will be insensitive to the wave's phase. This is the appropriate situation for a QND energy measurement. The error in the measurement (if other, in principle avoidable, uncertainties are absent) will be

$$\Delta E_{\text{measure}} = \frac{d \, \Delta P}{\tau} \, , \tag{12.3}$$

where ΔP is the initial uncertainty in the transverse component of the electron's momentum.

The electron's oscillations at frequency ω_e exert a back action on the waveguide. This back action is equivalent to a change in the guide's dielectric susceptibility by an amount proportional to the distance x between the guide and the electron. This change of susceptibility produces a corresponding x-dependent change

$$v = v_o(1 - x/d)$$

in the speed of light in the waveguide. Here $1/d$ is the same coupling constant as appears in the quadratic force on the electron. Any uncertainty Δx in the electron's transverse location will produce a corresponding uncertainty

$$\Delta v = v_o \frac{\Delta x}{d}$$

in the wave's propagation speed, a corresponding uncertainty

$$\Delta z = \tau \Delta v = \tau v_o \frac{\Delta x}{d}$$

in the location of the wave packet after the interaction is finished, and hence an uncertainty

$$\Delta \tau_{\text{perturb}} = \frac{\Delta z}{v_o} = \tau \frac{\Delta x}{d} \tag{12.4}$$

in the time at which the wave packet arrives at any given location beyond the interaction region. The product of the right-hand sides of equations (12.3) and (12.4) gives, in view of

$$\Delta x \cdot \Delta P \geq \frac{\hbar}{2} \, ,$$

an uncertainty relation constraining the QND measurement of the traveling wave's energy:

$$\Delta E_{\text{measure}} \cdot \Delta \tau_{\text{perturb}} \geq \frac{\hbar}{2} \ . \tag{12.5}$$

Notice that the uncertainty $\Delta \tau_{\text{perturb}}$ is equivalent to an effective broadening of the electromagnetic wave packet by the amount $\Delta \tau_{\text{perturb}} v_o = \Delta z$.

An analogous analysis can be carried out for any other scheme for a QND measurement of a traveling wave's energy. The only things that will change in the fundamental formulae are the meanings of such quantities as x and P. For example, in a scheme based on a cubic dielectric nonlinearity, the role of P is played by the phase of the probe wave and the role of x is played by the probe wave's energy.

12.3 Frequency-anticorrelated quantum state

For a mode of an electromagnetic resonator, a measurement of the oscillation energy is equivalent to a measurement of the number of quanta because the energy of each quantum, $\hbar\omega$, is precisely known (the mode's eigenfrequency ω can be measured in advance with arbitrary precision). Not so for a traveling wave: in a wave packet of duration τ, the frequency of each photon can be defined only up to an accuracy $\Delta\omega = 1/2\tau$. As a result, each photon has an energy uncertainty

$$\Delta E_1 = \hbar \Delta\omega = \frac{\hbar}{2\tau} \ . \tag{12.6}$$

Now, suppose that a perfect QND measurement has been made on the energy of some mode of a resonator, and thus the number of photons in the resonator is known precisely. Suppose, further, that after this measurement the resonator is coupled to a waveguide, and the photons thereby leak out. After some time τ (that depends on the strength of the coupling), *all* the photons will have left the resonator and will be congregated into a wave packet of duration τ that contains a precisely defined number n of photons.

The coupling between the resonator and the waveguide is a parametric process: although it does not create or destroy photons, it randomly changes the photons' frequencies. As a result, each photon in the wave packet has an energy uncertainty $\Delta E_1 = \hbar/2\tau$ [equation (12.6)]. Because the uncertainties for different photons are independent, the uncertainty of the packet's total energy will be $\sqrt{n}$ times that of the individual photons:

$$\Delta E_n = \sqrt{n} \cdot \Delta E_1 = \frac{\hbar}{2\tau}\sqrt{n} \ . \tag{12.7}$$

Suppose, now, that we measure the packet's total energy by the method of the last section. According to equation (12.5) this measurement lengthens the wave packet. However, the energy can be measured sufficiently precisely to strongly beat the standard quantum limit:

$$\Delta E \ll \frac{\hbar}{2\tau'}\sqrt{n} \ , \qquad (12.8)$$

where

$$\tau' = \tau + \Delta\tau_{perturb}$$

is the new duration of the wave packet. If $\Delta\tau_{perturb} \gg \tau$, then the measurement can achieve

$$\Delta E = \frac{\hbar}{2\tau'}$$

(the same as for a monophotonic state of the same duration).

This accurate measurement of the wave packet's energy drives it into a new quantum state. The inequality (12.8) for this quantum state implies that the deviations of the frequencies of individual quanta from their mean value

$$\tilde{\omega} = \frac{\tilde{E}}{n\,\hbar\omega}$$

(where $\tilde{E}$ is the wave packet's measured total energy) are no longer independent. The individual frequencies are now, one can show, anticorrelated random variables, and correspondingly the new quantum state is called the "frequency-anticorrelated state."[66]

This state is an analog of the photon antibunched state of section 12.1. In the antibunched state, the spatial distribution of photons is more uniform than for a random flux of particles, and this reduced randomness can be detected in the statistics of the photons' arrival times at the detector. In the frequency-anticorrelated state, there is a certain ordering of the photons by their frequencies. To detect this effect one must use a spectrograph-type measuring device. In the absence of any frequency correlations, after registering a photon with an energy higher than the mean, one cannot say anything about the sign of the frequency deviation of the next photon. In the frequency-anticorrelated state, there is an increased probability for the second photon to have an energy below the mean.

It is incorrect to presume that, when one measures the total energy of the packet, the photons somehow are forced to "line up" according to their frequencies. Instead, the "true" mean frequency of the wave packet becomes known, and the deviations of the frequencies of individual photons relative to it become anticorrelated.

12.4 Doppler measurements with frequency-anticorrelated photons

In section 1.4 we analyzed a Doppler-based device for measuring the speed of a mechanical object—a device originally proposed by John von Neumann. This measuring device permits one, in principle, to measure the speed with arbitrarily high precision. However, simple estimates [based on equation (1.15)] show that, to obtain high precision, one must use a photon of very high energy.

Instead of a single photon, it should be possible to use a multiphoton wave packet. When a group of photons is reflected from the object, whose initial speed is v, the group's total energy changes from $\boldsymbol{E}$ to

$$\boldsymbol{E'} = \boldsymbol{E}\frac{1 - v/c}{1 + \dfrac{v}{c} + \dfrac{2\boldsymbol{E}}{mc^2}\left[1 - \dfrac{v^2}{c^2}\right]^{1/2}} \, ,$$

where c is the speed of light and m is the object's mass. When viewed as quantum mechanical operators, all the quantities in this formula commute with each other. This means that the formula maintains its same form in the quantum case, with all the classical quantities replaced by their corresponding operators.

If we know the photons' total energy before and after the measurement, then from the difference $\boldsymbol{E'} - \boldsymbol{E}$ we can infer the object's speed v. The energy $\boldsymbol{E'}$ after the reflection can be measured with arbitrarily high precision by absorbing it in some detector. Then the error Δv in the inferred speed will be determined by the initial energy uncertainty $\Delta \boldsymbol{E}$. In the nonrelativistic approximation ($v \ll c$ and $\boldsymbol{E} \ll mc^2$),

$$\Delta v = \frac{c}{2}\frac{\Delta \boldsymbol{E}}{\boldsymbol{E}} \, . \tag{12.9}$$

It is important to emphasize the difference between this measurement procedure and the traditional procedure for Doppler measurements of the speed: here the photons' total energy is measured; in the traditional procedure their frequency is measured using a frequency-sensitivity receiver. It is known[67] that the precision of the traditional procedure is constrained by the standard quantum limit

$$\Delta v_{\mathrm{SQL}} = \sqrt{\frac{\hbar}{2m\,\tau_{\mathrm{measure}}}} \, ,$$

where τ_{measure} is the measurement time and m is the mass of the measured object. Not so for the present procedure, as the following analysis will show.

Let us begin our analysis with the simplest case of a coherent initial state for the wave packet. In this case $\Delta E = \hbar\omega\sqrt{n}$, where n is the mean number of quanta, so

$$\Delta v = \frac{c}{2\sqrt{n}} \ .$$

For a wave packet with mean wavelength 6×10^{-5} cm and with total energy 10^7 erg $= 1$ joule, the ultimate precision is very poor: $\Delta v \simeq 10$ cm/sec.

The precision can be substantially better if the number of photons in the wave packet is known exactly. The uncertainty of the packet's total energy is then given by equation (12.7), and insertion of this uncertainty into (12.9) gives

$$\Delta v = \frac{c}{4\omega\tau\sqrt{n}} \ .$$

If $\tau = 10^{-3}$ sec and the other parameters are the same as in the previous example, then $\Delta v \simeq 1.5\times10^{-12}$ cm/sec. This precision is very high; it exceeds the standard quantum limit for speed of $m < 0.1$ gram. However, in some situations (for example, in gravitational-wave experiments which involve masses much larger than 0.1 gram), even higher precision is needed. In such cases the frequency-anticorrelated states may be useful because for a definite duration τ of the wave packet, it has the minimum possible energy uncertainty (12.6). Using this state, we obtain the minimum error

$$\Delta v = \frac{c}{4\omega\tau n} = \frac{\hbar c}{2E\tau} \ . \tag{12.10}$$

For the same values of the parameters as was used above, this gives $\Delta v \simeq 0.7\times10^{-21}$ cm/sec.

In closing this section, let us discuss the influence of dissipation on the measurement precision. Denote by R the probability for a photon to reach the detector without being absorbed in either the waveguide or the reflector. There are two mechanisms by which the dissipation can produce uncertainties in the total energy of the group of photons. First, the number of absorbed photons is a random quantity with variance

$$(\Delta_R n)^2 = nR(1-R) \simeq 10^{-5}.$$

Second, the energy of each of the $n(1-R)$ absorbed photons has an uncertainty of $\hbar/2\tau$. Therefore, the net uncertainty produced by the dissipation will be

$$(\Delta_R \boldsymbol{E})^2 = (\hbar\omega\Delta_R n)^2 + n(1-R)\left[\frac{\hbar}{2\tau}\right]^2 = \hbar^2 n(1-R)\left[\omega^2 R + \frac{1}{(2\tau)^2}\right]. \quad (12.11)$$

Since always $\omega\tau \gg 1$, the first term gives the dominant contribution. Inserting this energy uncertainty into equation (12.9), we obtain for the contribution of the dissipation to the error in the inferred speed of the object

$$\Delta_R v = \frac{c}{2}\sqrt{\frac{1-R}{nR}}.$$

Even for the highest reflectivity now available at optical frequencies, $(1-R) \simeq 10^{-5}$ this gives a very low measurement precision ($\Delta v \simeq 0.1$ cm/sec if $\boldsymbol{E} \simeq 10^7$ erg).

However, by changing the measurement scheme somewhat, one can improve the error substantially. The offending energy uncertainty $\Delta_R \boldsymbol{E}$ is produced by a random change in the number of photons, while the signal (of Doppler origin) is produced by a frequency-induced change in the energies of all the photons. If we measure not only the total initial and final energies, but also the number of photons in the wave packet, then it is possible to determine the number of photons that were absorbed and thereby eliminate the first term in equation (12.11). The error in the speed measurement in this case is

$$\Delta_R v = \frac{c}{4\omega\tau}\sqrt{\frac{1-R}{n}}.$$

This formula implies that the minimum allowed error (12.10) can be obtained only if the level of dissipation is so low that the mean number of photons absorbed is less than unity: $n(1-R) < 1$. However, if the mean number of absorbed photons is larger than one, by using a frequency-anticorrelated state, one can gain a factor $\sqrt{1-R}$ relative to the previous case.

A practical realization of this method of measuring speed would be a very difficult task—so difficult that the above analysis should be regarded as only tutorial.

Reference 68 presents another, perhaps more realistic variant of a QND speed meter. This scheme is based on two successive measurements of the object's position, and a direct subtraction of them inside the quantum probe in such a way that the observer obtains only information about their difference (and thus about the object's speed), and no information about the object's absolute position.

12.5* Statistical properties of a wave packet with a definite number of quanta

In this section we shall analyze the statistical properties of a wave packet that contains a definite number of quanta. We shall idealize the packet as propagating in an infinitely long, one-dimensional waveguide. This permits us to avoid all the technical difficulties associated with the vector nature of the electromagnetic field in three-dimensional space, while still preserving the principal features that are relevant to the theory of measurement.

Denote by $\hat{a}^{\dagger}(\omega)$ and $\hat{a}(\omega)$ the creation and annihilation operators for quanta propagating in the $+x$ direction. These operators obey the standard commutation relations for a continuous spectrum of modes:

$$[\hat{a}(\omega), \hat{a}^{\dagger}(\omega')] = \delta(\omega - \omega') . \tag{12.12}$$

Any quantum state that contains precisely n photons propagating in the $+x$ direction can be expressed as

$$|n> = \frac{1}{\sqrt{n!}} \int_0^{\infty} \psi(\omega_1,...,\omega_n) \hat{a}^{\dagger}(\omega_1) \cdots \hat{a}^{\dagger}(\omega_n) d\omega_1 \cdots d\omega_n |0> , \tag{12.13}$$

where $|0>$ is the waveguide's vacuum state, and the wave function $\psi(...)$ is normalized to unity:

$$\int_0^{\infty} |\psi(\omega_1,...,\omega_n)|^2 d\omega_1 \cdots d\omega_n = 1$$

and is symmetric with respect to all pairs of arguments.

Let us try to find the minimum allowed uncertainty of the total energy

$$\hat{E} = \int_0^{\infty} \hbar\omega \hat{a}^{\dagger}(\omega) \hat{a}(\omega) d\omega$$

in this state, when the length of the wave packet is known. Using the commutator (12.12), one can show that the operator $\hat{E}$ acts on the state (12.13) in the following way:

$$\hat{E}|n>$$

$$= \frac{1}{\sqrt{n!}} \int_0^{\infty} (\hbar\omega_1 + ... + \hbar\omega_n) \psi(\omega_1,...,\omega_n) \hat{a}^{\dagger}(\omega_1) \cdots \hat{a}^{\dagger}(\omega_n) d\omega_1 \cdots d\omega_n |0> .$$

From this equation it follows that the mean energy is

$$<n|\hat{E}|n> = \int_0^{\infty} (\hbar\omega_1 + ... + \hbar\omega_n) |\psi(\omega_1,...,\omega_n)|^2 d\omega_1 \cdots d\omega_n ,$$

and its mean square energy is

$$\langle n \,|\, \hat{E}^2 \,|\, n \rangle = \int_0^\infty (\hbar\omega_1 + ... + \hbar\omega_n)^2 |\,\psi(\omega_1,...,\omega_n)\,|^2 d\omega_1 \cdots d\omega_n \ .$$

(The derivation of these relations uses the commutator (12.12) once more and the symmetry of the wave function.) The variance of the energy is defined in terms of the above two quantities by

$$(\Delta E)^2 = \langle n \,|\, \hat{E}^2 \,|\, n \rangle - \langle n \,|\, \hat{E} \,|\, n \rangle^2 \ . \tag{12.14}$$

The spatial distribution of the energy is characterized by the probability density

$$W(x) = \frac{\langle n \,|\, \hat{\rho}(x) \,|\, n \rangle}{\langle n \,|\, \hat{E} \,|\, n \rangle} \ , \tag{12.15}$$

where

$$\hat{\rho}(x) = \frac{\hbar}{2\pi c} \int_0^\infty \{\hat{a}^\dagger(\omega)\hat{a}(\omega')e^{i(\omega-\omega')t}$$

$$+ \frac{1}{2}[\hat{a}(\omega)\hat{a}(\omega')e^{-i(\omega+\omega')t} + \hat{a}^\dagger(\omega)\hat{a}^\dagger(\omega')e^{i(\omega+\omega')t}]\} d\omega d\omega'$$

is the energy-density operator in the waveguide, and c is the speed of light. The probability distribution $W(x)$ is normalized to unity:

$$\int_{-\infty}^{+\infty} W(x)dx \equiv 1 \ .$$

The variance of this probability distribution, denoted $(\tfrac{1}{2}c\,\tau)^2$, defines the wave packet's duration τ and length $c\,\tau$. By inserting expression (12.12) into equation (12.15) and performing a rather long calculation, we obtain

$$\left[\frac{\tau}{2}\right]^2 = \frac{\hbar n}{\langle E \rangle} \int_0^\infty |(\partial/\partial\omega_1)[\sqrt{\omega_1}\cdot\psi(\omega_1,...,\omega_n)]|^2 d\omega_1 \cdots d\omega_n \ . \tag{12.16}$$

When the wave packet's bandwidth is much narrower than its mean frequency $\bar{\omega}$ (the case of greatest practical interest), the product $\Delta E \cdot \tau$ is minimized by a Gaussian wave function:

$$\psi(\omega_1,...,\omega_n) = \psi_o \exp[-\frac{1}{4}\sum_{j,k=1}^n \Lambda_{jk}(\omega_j - \bar{\omega})(\omega_k - \bar{\omega})] \ . \tag{12.17}$$

The coefficients Λ_{jk} appearing here form a matrix whose inverse, $||B_{jk}|| \equiv ||\Lambda_{jk}||^{-1}$, is formed from the second moments of the probability distribution

$$W(\omega_1,...,\omega_n) = |\psi(\omega_1,...,\omega_n)|^2$$

for the photon frequencies.

From the symmetry of the wave function it follows that all the diagonal elements of Λ_{ij} are equal to each other, as are all the off-diagonal elements; and the same must then be true for B_{ij}. Thus, we can write the second moments of the photons' frequency distribution in the form

$$B_{jk} = \sigma^2[\delta_{jk}+\beta(1-\delta_{jk})] ,\qquad (12.18)$$

where σ^2 is the variance of the distribution and β is its correlation coefficient. By inserting the wave function (12.17) into equations (12.14) and (12.16), taking account of (12.18), we obtain

$$(\Delta E)^2 = (\hbar\sigma)^2 n[(n-1)\beta+1] , \qquad \tau^2 = \frac{1}{\sigma^2}\frac{(n-2)\beta+1}{(1-\beta)[(n-1)\beta+1]} .$$

If $\beta = 0$, i.e. when there is no correlation of the frequencies,

$$\Delta E = \hbar\sigma\sqrt{n} , \quad \tau = \frac{1}{\sigma} ,$$

and therefore

$$\Delta E\cdot\tau = \hbar\sqrt{n} .$$

The quantity σ in this case is evidently the frequency uncertainty of each individual photon. If $\beta > 0$, the photon frequencies are correlated; if $\beta < 0$, they are anticorrelated. The minimum possible value of β is

$$\beta_{min} = -\frac{1}{n-1} .$$

When $\beta\to\beta_{min}$, then $\Delta E\to 0$ and $\tau\to\infty$, but their product remains finite:

$$\Delta E\cdot\tau = \hbar .$$

From a comparison of this formula with equation (12.5) it is evident that, if one makes a sufficiently precise measurement of the energy of a traveling electromagnetic wave, the wave's photon frequencies must become anticorrelated.

Conclusion

The background for this book was a shift of interest among experimental physicists that began approximately 20 years ago. The shift was from quantum measurements on large ensembles of microscopic quantum objects, to experiments of very high precision on single, macroscopic objects. This shift was accompanied by developments in the "culture" of such measurements and the invention and elaboration of new experimental methods. In parallel with these experimental developments, there was a noticeable increase of theoretical interest in the foundations and interpretation of quantum theory, and even in attempts to revise its foundational principles. Although these efforts have not produced any significant changes in the fundamental structure of quantum theory, they have revealed a rather large number of "blank spots" and unsolved problems, to which the founders of quantum physics had paid no attention. A reasonable number of the blank spots have been filled in during the past decade.

The reader has probably noticed that this book can be divided into two parts. In the first (chapters I—VII and XI), the authors have tried to present the modern approach to the quantum theory of measurement. This part can be regarded as a tutorial appendix to textbooks for advanced undergraduate or graduate-level courses in quantum mechanics, where little attention is usually paid to measurement theory. The second part is a short review of new methods of quantum measurement for macroscopic objects. The authors have made no attempt to include here all the problems and directions of recent research, since this part of the book is likely to become obsolete soon. The selection of topics here was based mainly on the personal interests of the authors. The main goal in this part of the book was to illustrate the general ideas from the first part by examples of concrete measuring devices. Precisely because of this the authors have limited themselves to a short description of such important topics as the properties of squeezed states and techniques for generating and detecting them, and have totally omitted other important topics such as: (i) the use of squeezed states in information transfer; (ii) very interesting experiments on single microscopic quantum objects (e.g., the single atom maser,[69] and an atom trapped in laser beams[70]); (iii) the ''quantum computer'' (the possibility of quantum mechanical calculations performed without the absorption of energy, i.e. reversible calculations[71, 74]), which is closely connected to QND measurements of the number of quanta; and (iv) quantum measurements in strong gravitational fields and in accelerated reference frames.[75-78] However, in future developments of these omitted topics, the ideas presented in the first part of the book may be useful.

It is worth noting that as yet there is no completely satisfactory theory of nonlinear quantum measurements. The authors hope that this book will be useful for those who try to construct such a theory.

One of the authors (V.B.) expresses his gratitude to the California Institute of Technology for a three-month appointment as a Sherman Fairchild Distinguished Scholar, which gave him an opportunity to work with an outstanding team of physicists and also provided excellent conditions for preparing the final draft of this manuscript.

Both authors are indebted to the editor, Professor Kip S. Thorne, whose attitude toward the editing of this book was one of the best known. It is appropriate to note here that he himself has made major contributions to the field of physics covered by this book. The authors also express their sincere gratitude to Professor Carolee Winstein, whose help, advice, and inspiration in the process of editing were priceless.

References

[1] M. Planck, *Warmentrahlung* (Leipzig, 1906).

[2] A. Einstein, Ann. Physik, [4], Bd. 17 (1905).

[3] P.N. Lebedev, Journal of Russian Physical-Chemical Society, **33**, 53 (1901); in Russian.

[4] A.J. Dempster and H.F. Batho, Phys. Rev., **30**, 644 (1927).

[5] Y.R. Shen, *The Principles of Nonlinear Optics*, (Wiley, New York, 1984).

[6] C.K. Hong and L. Mandel, Phys. Rev. Lett., **56**, 58 (1986).

[7] D.N. Klyshko, Dov. J. Quantum Electronics, **7**, 591 (1977).

[8] C.J. Davisson and L.H. Germer, Phys. Rev., **30**, 705 (1927).

[9] W. Heisenberg, Z. Physik, Bd. 4, 879 (1925).

[10] E. Schrödinger, Ann. d. Physik, [4], Bd. 79, 361; Bd. 80, 437; Bd. 81, 109 (1926).

[11] P.A.M. Dirac, *The Principles of Quantum Mechanics* (Clarendon Press, Oxford, 1930).

[12] J. von Neumann, *Matematische Grundlagen der Quantenmechanik* (Berlin, 1932).

[13] N. Bohr, Phys. Rev., **48**, 696 (1935).

[14] A. Einstein, in *Albert Einstein: Philosopher-Scientist*, edited by P. Schilpp (Tudor, New York, 1949).

[15] V.B. Braginsky, *Physical Experiments with Test Bodies*, NASA Technical Translation TT F-672 (U.S. Technical Information Service, Springfield, VA; 1972).

[16] V.B. Braginsky and Yu I. Vorontsov, Sov. Phys.–Uspekhi, **17**, 644 (1975).

[17] Kh.S. Bagdasarov, V.B. Braginsky, V.B. Mitrofanov, and V.S. Shiyan, Bull. Moscow State University, Ser. 3, No. 9, p. 98 (1977); in Russian.

[18] V.B. Braginsky and V.P. Mitrofanov, *Systems with Small Dissipation*

(University of Chicago Press, Chicago, 1985).

[19] L. de Broglie, Comptes Rend. **177**, 517 (1923).

[20] A. Compton and D. Simons, Phys. Rev., **26** (1925).

[21] A. Einstein, B. Podolsky, and N. Rosen, Phys. Rev., **47**, 777 (1935).

[22] A.A. Kurishka, *Quantum Optics and Optical Location* (Soviet Radio Pbl., Moscow, 1973); in Russian.

[23] R.L. Stratonovitch, J. Stochastics, **1**, 87 (1973).

[24] C.W. Helstrom, *Quantum Detection and Estimation Theory* (Academic Press, New York, 1976).

[25] A.S. Kholevo, *Probabilistic and Statistical Aspects of Quantum Theory* (North Holland, Amsterdam, 1982).

[26] L.I. Mandelstam, *Lectures on Optics, the Theory of Relativity, and Quantum Mechanics* (Nauka, Moscow, 1972); in Russian

[27] V.B. Braginsky, Yu.I. Vorontsov, and F.Ya. Khalili, Sov. Phys.–JETP, **46**, 705 (1977); W. Unruh, Phys. Rev. D, **18**, 1764 (1978).

[28] M.D. Levenson, R.M. Shelby, M. Reid, and D.F. Walls, Phys. Rev. Lett., **57**, 2473 (1986).

[29] Yu.I. Vorontsov, *Theory and Methods of Macroscopic Measurements* (Nauka, Moscow, 1989); in Russian.

[30] M.B. Mensky, *The Group of Paths: Measurements, Fields, Particles* (Nauka, Moscow, 1983); in Russian. C.M. Caves, Phys. Rev. D, **33**, 1643 (1986) and **35**, 1815 (1987).

[31] C.M. Caves, Phys. Rev. D, **26**, 1817 (1982).

[32] Yu.I. Vorontsov and F.Ya. Khalili, Radiotechnica i. Electronica, **27**, 2392 (1982); in Russian.

[33] W.H. Zurek, in *The Wave-Particle Dualism*, proceedings of a symposium in honor of the 90th birthday of Louis de Broglie, edited by S. Diner, D. Fargue, G. Lochak, and F. Selleri (Reidel, Dordrecht, 1984) p. 515.

[34] L.A. Khalfin, Sov. Phys.–JETP, **6**, 1053 (1958). B. Misra and E.C.G. Sudarshan, J. Math. Phys., **18**, 756 (1977).

[35] W.M. Itano, D.J. Heinzen, J.J. Bollinger, and D.J. Wineland, Phys. Rev. A, **41**, 2295 (1990).

[36] V.B. Braginsky, Sov. Phys.–Uspekhi, **156**, 93 (1988).

[37] Yu.I. Vorontsov and F.Ya. Khalili, Sov. Phys.–JETP, **55**, 43 (1982).

[38] V.B. Braginsky, Yu.I. Vorontsov, and F.Ya. Khalili, Sov. Phys.–JETP Lett., **33**, 405 (1981).

[39] K.S. Thorne, R.W.P. Drever, C.M. Caves, M. Zimmermann, and V.D. Sandberg, Phys. Rev. Lett., **40**, 667 (1978). C.M. Caves, K.S. Thorne, R.W.P. Drever, V.D. Sandberg, and M. Zimmermann, Rev. Mod. Phys., **52**, 341 (1980).

[40] K.S. Thorne, in *300 Years of Gravitation*, eds. S.W. Hawking and W. Israel (Cambridge Univ. Press, Cambridge, 1988).

[41] R.D. Reasenberg, Phil. Trans., **310**, 227 (1984).

[42] P. Bender, Report at the 27th COSPAR Congress (1988); edited by R. Reasenberg and R.F.C. Vessot.

[43] J Hough, B.J. Meers, G.P. Newton, N.A. Robertson, H. Ward, G. Leuchs, T.M. Niebauer, A. Rüdiger, R. Schilling, L. Schnupp, H. Walther, W. Winkler, B.F. Schutz, J. Ehlers, P. Kafka, G. Schäfer, M.W. Hamilton, I. Schütz, H. Welling, J.R.J. Bennett, I.F. Corbett, B.W.H. Edwards, R.J.S. Greenhalgh, and V. Kose, *Proposal for a Joint German-British Interferometric Gravitational-Wave Detector* (Max Planck Institute fur Quantenoptic, Garching, Germany, MPQ-147, 1989).

[44] D. Shoemaker, R. Schilling, L. Schnupp, W. Winkler, K. Maischberger, and A. Rudiger, Phys. Rev. D, **38**, 423 (1988).

[45] V.B. Braginsky, V.I. Panov, and V.D. Popel'nyuk, Sov. Phys—JETP Lett., **33**, 405 (1981).

[46] V.I. Panov and A.A. Sobyanin, Sov. Phys.—JETP Lett., **35**, 404 (1982).

[47] V.B. Braginsky and S.P. Vyatchanin, Sov. Phys.—JETP, **47**, 433 (1978). V.B. Braginsky, S.P. Vyatchanin and V.I. Panov, *Sov. Phys.—Doklady,* **24**, 562 (1979).

[48] W.E. Lamb, in *Quantum Optics and Electronics*, (Sci. Publ, 1965).

[49] V.V. Migulin, V.I. Medvedev, E.P. Mustel', and V.N. Parigin, *Foundations of the Theory of Oscillations* (Nauka, Moscow, 1978); in Russian.

[50] V.B. Braginsky, M.L. Gorodetsky, and V.S. Il'chenko, Phys. Lett. A, **137**, 393 (1989).

[51] V.B. Braginsky and S.P. Vyatchanin, Sov. Phys.—Doklady, **26**, 686 (1981).

[52] G.J. Milburn and D. Walls, Phys. Rev. A, **28**, 2065, 1983.

[53] V.B. Braginsky, V.S. Il'chenko, and M.L. Gorodetsky, Sov. Phys.—Uspekhi, in press [Russian is **160**, 157 (1990)]/

[54] V.B. Braginsky and F.Ya. Khalili, Sov. Phys.—JETP, **51**, 859 (1980).

[55] R. Hanbury-Brown and R.Q. Twiss, Nature, **177**, 27 (1956).

[56] R.J. Glauber, in *Quantum Optics and Electronics*, (Sci. Publ., 1965).

[57] V.B. Braginsky and F.Ya. Khalili, Sov. Phys.—JETP, **57**, 1124 (1983).

[58] Y. Yamamoto and H.A. Haus, Rev. Mod. Phys., **158**, 1001, 1986.

[59] C.M. Caves, Phys. Rev. D, **23**, 1693, 1981.

[60] D.F. Walls and S.E. Slusher, eds., J. Opt. Soc. Am., **B4**, special issue (October 1987).

[61] R. Loudon and P.L. Knight, eds., J. Mod. Opt., special issue (June 1987).

[62] Y. Yamamoto, S. Mashida, and O. Nillson, in *Coherence, Amplification, and Quantum Effects in Semiconductor Lasers*, edited by Y. Yamamoto (Wiley, New York, 1989).

[63] K. Watanabe, H. Nakano, A. Honold, and Y. Yamamoto, Phys. Rev. Lett. A, **132**, 206 (1988).

[64] V.B. Braginsky and S.P. Vyatchanin, Phys. Lett. A, **132**, 206 (1988).

[65] A. Gaponov and M.A. Miller, Sov. Phys.—JETP, **7**, 168 (1958).

[66] V.B. Braginsky and F.Ya. Khalili, Sov. Phys.—JETP, **67**, 84 (1988).

[67] Yu.I. Vorontsov and T.P. Bocharova, Sov. Phys.—JETP, **57**, 933 (1983).

[68] V.B. Braginsky and F.Ya. Khalili, Phys. Lett. A, **147**, 251 (1990).

[69] J. Krause, M.O. Scully and H. Walther, Phys. Rev. A, **36**, 4547 (1987). G. Rempe, W. Scheeich, M.O. Scully, and H. Walther, in Proceedings of the 3rd international symposium on *Foundations of Quantum Mechanics in the Light of New Technology* (Tokyo, 1990).

[70] J.W.R. Tabosa, G. Chen, Z. Hu, R.B. Lee, and H.J. Kimble, Phys. Rev. Lett., **66**, 3245 (1991).

[71] R.P. Feynman, Foundations of Physics, **16**, 507 (1986). E. Fredkin and T. Toffoli, Int. J. Theor. Phys., **21**, 219 (1982).

[72] W.G. Unruh, Phys. Rev. D **19**, 2888 (1979). C.M. Caves, section IV of the second of Refs. 39.

[73] A. Sudbury, *Quantum mechanics and the particles of nature*, Chapter 5 (Cambridge Univ. Press, Cambridge, 1982).

[74] C.H. Bennett, Int. J. of Theor. Phys., **21**, 905 (1982).

[75] S.W. Hawking, Nature, **248**, 30 (1974).

[76] W.G. Unruh, Phys. Rev. D, **14**, 870 (1976).

[77] N.D. Birrel, P.C.W. Davies, *Quantum fields in curved space* (Cambridge Univ. Press, Cambridge, 1982).

[78] W.G. Unruh, R.W. Walls, Phys. Rev. D, **29**, 1047 (1984).

Subject Index

Printed in the United States
By Bookmasters